Next-Gen Manufacturing 2022

Topical Conference at the 2022 AIChE Annual Meeting

Phoenix, Arizona, USA
13-18 November 2022

ISBN: 978-1-7138-7904-6

Printed from e-media with permission by:

Curran Associates, Inc.
57 Morehouse Lane
Red Hook, NY 12571

CURRAN ASSOCIATES INC. proceedings.com

Some format issues inherent in the e-media version may also appear in this print version.

AIChE
120 Wall Street, FL 23
New York, NY 10005-4020

Phone: (800) 242-4363
Fax: (203) 775-5177

www.aiche.org

Additional copies of this publication are available from:

Curran Associates, Inc.
57 Morehouse Lane
Red Hook, NY 12571 USA
Phone: 845-758-0400
Fax: 845-758-2633
Email: curran@proceedings.com
Web: www.proceedings.com

TABLE OF CONTENTS

3D PRINTING OF FUNCTIONAL MATERIALS

NEXT-GEN MANUFACTURING IN CHEMICAL AND ENERGY SYSTEMS

CYBERSECURITY AND HIGH-PERFORMANCE COMPUTING IN NEXT-GEN MANUFACTURING

FUTURE OF MANUFACTURING AND EMERGING TECHNOLOGIES

INDUSTRIAL INTERNET OF THINGS (IIOT), SMART AND SOFT SENSORS IN PROCESS MANUFACTURING AND BEYOND

INDUSTRY 4.0, DIGITAL TWINS, AND DIGITAL TRANSFORMATION

APPLIED ARTIFICIAL INTELLIGENCE, BIG DATA, AND DATA ANALYTICS METHODS FOR NEXT-GEN MANUFACTURING EFFICIENCY II

NEXT-GEN MANUFACTURING IN PHARMA, FOOD, AND BIOPROCESSING II

MODELING, OPTIMIZATION, AND CONTROL IN NEXT-GEN MANUFACTURING I

POLYMERS IN ADDITIVE MANUFACTURING

APPLIED ARTIFICIAL INTELLIGENCE, BIG DATA, AND DATA ANALYTICS METHODS FOR NEXT-GEN MANUFACTURING EFFICIENCY I

NEXT-GEN MANUFACTURING IN PHARMA, FOOD, AND BIOPROCESSING I

MODELING, OPTIMIZATION, AND CONTROL IN NEXT-GEN MANUFACTURING II

38a 3D printing all-aromatic polyimides with light: Photoreactive supramolecular polymeric salts as a versatile printing platform

Timothy E. Long, Christopher B. Williams

Future advances in additive manufacturing (AM) will require the rethinking and redesign of traditional macromolecular structure. Advanced macromolecular materials for advanced manufacturing require a precisely tailored balance of reactivity and rheological performance that collectively ensure precise resolution from diverse additive manufacturing modalities. Moreover, research must impose a lens of sustainability to ensure next-generation advanced materials for advanced manufacturing minimize negative consequences on the environment. Our recent research has focused on the printing of high molecular weight polymers using vat photopolymerization (VP) and UV-assisted direct ink write (UV-DIW) where printed organogel and hydrogel precursors enabled the printing of high-performance fully aromatic polyimides and diene-based copolymer elastomers, respectively. Our recent results illustrate the solvent-free stereolithographic printing of high molecular weight polymers as low viscosity aqueous colloids to prepare elastomers with unprecedented three-dimensional geometries. This lecture will focus on printed three-dimensional all-aromatic polyimide objects that present identical thermomechanical properties compared to polyimide films that are derived from legacy manufacturing processes. Acrylate and epoxide functionalities continue to play an important role in the design of photoreactive polymers for AM, however, recent adaptations of thiol-ene chemistry, photo-acid and photo-base generators, and unsaturated co-polyesters exemplify the rapidly expanding toolbox of functional polymers for additive manufacturing. The necessity to maintain a lens of materials sustainability will fuel the discovery of functional polymers that are amenable to recyclability and circularity, thus catalyzing the emerging field of sustainable additive manufacturing (SAM).

Supramolecular chemistry offers unique opportunities for the design of thermo-reversible polymers for additive manufacturing. Our earlier efforts demonstrated the importance of electrostatic interactions and multiple hydrogen bonding for fused filament fabrication of water- soluble polyesters and polyurethanes with improved isotropic properties. Most recently our research has focused on the formation of polymeric salts as precursors for vat photopolymerization (**Figure 1A**). Polymeric salts, commonly termed polysalts, that are derived from various photo-reactive dicarboxylic acids with various diamines are amenable to vat photopolymerization and with subsequent thermal processing to fully aromatic polyimides. This approach allows for low solution viscosities, which minimizes dipolar aprotic organic solvents and leads to reduced shrinkage in printed lattices. Recent efforts have expanded the library of suitable precursors with attention to the formation of more stable polymeric salts, thus allowing for longer shelf lives for polyimide precursors and more reproducible printing. This advance was based on an *in-situ* FTIR spectroscopic investigation, which demonstrated the importance of understanding aryl amine-containing monomer nucleophilicity in the presence of adjacent photo-reactive acrylate functionality. Avoiding the Michael addition reaction allowed for the formation of polymeric salts of amine-containing aromatic monomers with acrylate functionalities with a preservation of photo-reactivity. Current efforts involve an investigation of the role of crosslink density of the organogels on the printing process and structure of the intermediate high modulus green bodies.

Polymeric salts have also enabled pathways for the formation of porous printed objects with either the introduction of porogenic sequences as graft copolymers or facile solvent swapping technique followed with lyophilization to form aerogels (**Figure 1B**). Scanning electron microscopy (SEM) and small angle x-ray scattering (SAXS) revealed the formation of well-defined porous structures and demonstrated the opportunity to print porous polyimides with the preservation of geometric complexity. This lecture will also illustrate the opportunity to print monolithic carbon upon pyrolysis of the fully aromatic polyimide structures. The geometric complexity was preserved in the printed object upon pyrolysis, suggesting opportunities for printed catalysts, electrodes, and membranes. Raman spectroscopy confirmed the balance of disordered and graphitic carbon as a function of polyimide composition and processing conditions.

This lecture will introduce the concept of printing hydrogels and organogels with a focus on recent efforts to print all-aromatic polyimides. Success demanded facile and versatile synthetic strategies coupled with tailored post-process imidization strategies, representing the criticality of concurrent design of advanced materials and next-generation machines.

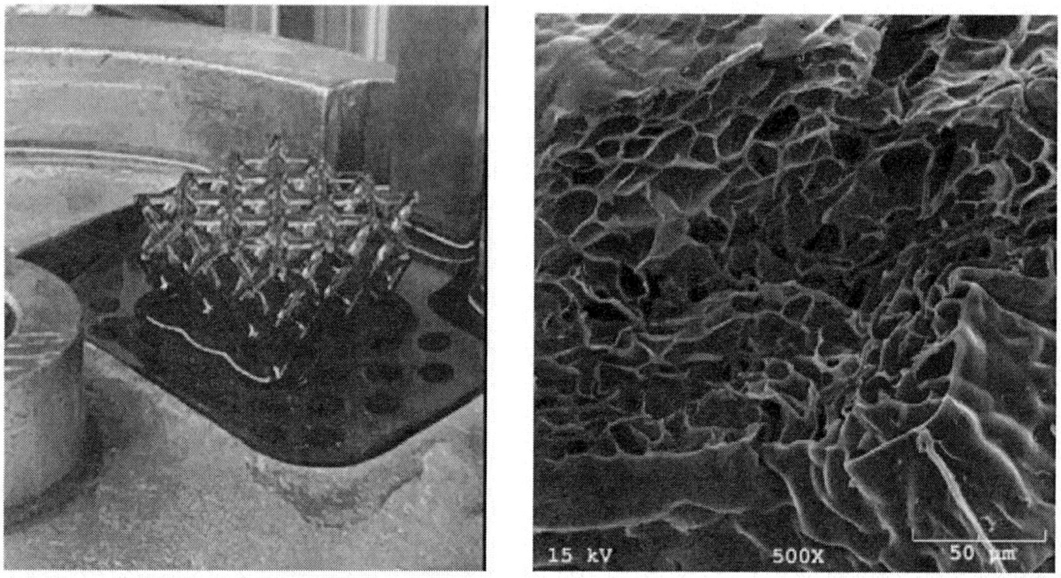

Figure 1. (left) Polyimide polymeric salts allow for the vat photopolymerization of latticved organogels that are amenable to imidization with preservation of complex, micron-scale objects, and **(right)** lyophilization of printed organogels provides printed polyimide aerogels with well defined porous structure as revealed with scanning electron microscopy.

38b Commercialization of Additive Manufacturing, Some Pitfalls and Consideration

Nima Yazdanpanah

Commercialization, technology transfer, and scale-up of Additive Manufacturing has its own challenges and required careful considerations, in addition to the general new technology commercialization challenges. These specific issues are due to the diversity of Additive Manufacturing, technology, and applications, technoeconomic, supporting platforms, and market demand. However, the enabling aspect of Additive Manufacturing could supersede some conventional technoeconomic issues.

This presentation will be in three sections: starting with the general commercialization issues and bringing a technology from lab to the market, with common pitfalls and some workarounds; Continuing with the specific challenges with the Additive Manufacturing domain and breaking down the steps with life cycle analysis; And at the end, will discuss some case studies.

38c Digital Light Processing of Highly-Filled Polymer Composites with Tailorable Mechanical Properties

Amy Peterson, Ye Wang

The effect of surface chemistry and interfacial interactions on the mechanical and thermal properties of additively manufactured composites were investigated using a high volume fraction glass microsphere-reinforced system. Interfacial interactions between the resin system and glass microspheres were found to dictate the overall performance of the composites. Better resin-glass wettability led to simultaneous and anomalous increases in stiffness, toughness, and strength. Interestingly, composites where glass was functionalized to covalently bond with the resin phase only showed moderate improvements in mechanical properties as compared to systems with high resin-glass wettability.

38d Directionally Dependent Fluid Behavior from Uniform Periodic Structures: Influence of Design and Additive Process Parameters

Ian Woodward, Catherine Fromen

Additive manufacturing has enabled the realization of constructs like periodic lattice structures, which are wide-ranging in their design and application. In a typical scenario, repeat units are patterned to create a macroscopic form for structural applications requiring isotropic or load-optimized performance. More recent chemical engineering applications include these additive-only lattice structures for their utility in heat exchange and process intensification, still largely based on symmetric, uniform repeat units. However, given the various processes and vast design space that additive manufacturing enables, there is an opportunity to take advantage of anisotropy that arises from the additive process or the parameters of the unit cell. Here we investigate the influences of design-stage and process-stage anisotropy on the fluidic performance of additively manufactured structures printed with the Carbon M1 platform. We examine a parametric array of pyramidal unit cell designs based on rotational symmetry, rather than reflection symmetry, and further consider the orientation of the unit cell with respect to print direction. Characterizing the dimensions of as-printed parts, we observe morphology variation based on part orientation and report the resulting effects of orientation and slicing dimension on fluidic behavior. In contrast to geometries with reflection symmetry about the plane normal to flow, pyramidal unit cells show statistically significant differences in pressure gradient in response to changing the direction of airflow across the part. As a result, we show that it is possible to combine the regularity of 3D printed lattices with controlled anisotropy to produce multifunctional parts with the potential to bolster the presence of additive manufacturing in applications relevant to chemical engineering.

38e Understanding and Characterising the Spreading of Cohesive Powders in Powder Bed Fusion Additive Manufacturing Via Discrete Element Modelling

Yi He, Ali Hassanpour, Andrew Bayly

The quality of the bed of powder that is spread during a powder bed fusion additive manufacturing process is important to quality of the part produced and the reliability of the process. The spread layer properties are determined by the process conditions, the machine set-up and geometry and the powder properties. The impact of the powder properties on the spread powder layer are not well understood due to limitations in both the numerical approaches used and the characterisation methods. This is particularly so for fine cohesive powders which require more complex contact models within the commonly used discrete element method. This study uses an efficient GPU based discrete element method with a novel methodology for scaling particle stiffness to increase simulation speed and maintain accuracy [1].

The validity of the method was verified for both packing and spreading of fine powders before the effect of powder cohesivity was investigated. To characterise the spread particle layer a number of alternative metrics were introduced to give insight into the layer structure, these included pore size distribution in the bed and powder cluster size distriburtions. The analysis showed that, for the conditions used, the packing density was relatively insensitive to cohesivity with particle Bond number spanning 0 to 200, however for very cohesive particles with Bo > than 200, a drop off in the bed density was seen. However significant differences in the structure of the spread layer were noted and several regimes identified. In particular, significant clustering, and less uniform spreading, was observed with lower cohesivity, flowable, powders. The mechanisms behind the observations are discussed and a dimensionless inertial number proposed to help interpret the phenomena observed.

The method provides a valuable tool and metrics to quantitatively evaluate the quality of a spread powder layers, but also enables a better understanding of the physics underlying the spreading of cohesive fine powders.

[1] Y. He et al., 2021, Applied Mathematical Modelling 90, 817-844

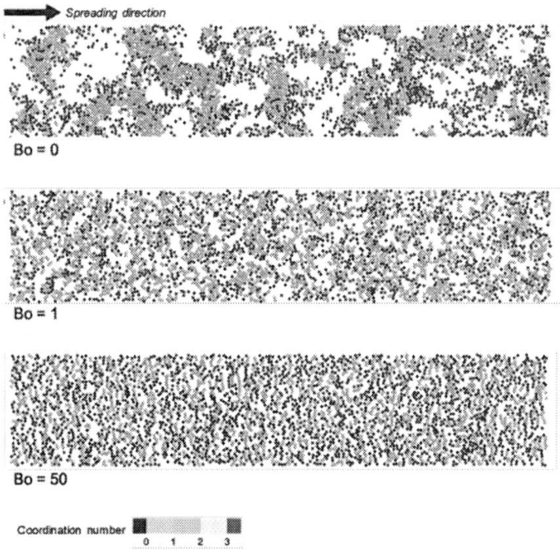

Spread layer seen from above with a range of cohesivities. Particles colored by co-ordination number

38f Improved Interlayer Adhesion in Fused Deposition Modeling (FDM) Printed Parts

Daniyal Shoukat, Jordan Totten, Jay Park, Nese Orbey, J Carson Meredith

Fused Deposition Modeling (FDM) is one of the most used Additive Manufacturing (AM) methods, mainly because of its low cost and minimal pre and post processing requirements. However, FDM printed parts exhibit poor mechanical properties, which result from the existence of porosity and a residual stress build-up because of rapid heating and cooling cycles between the extruder and the previously deposited layers. Consequently, FDM printed parts show poor interlayer adhesion, especially in the Z-orientation, leading to problems such as delamination and dimensional instability, greatly reducing the print quality. Insufficient adhesion between successive layers renders FDM printed parts much weaker than the traditional injection molded parts.

In this study, the effect of printing parameters, namely nozzle temperature, layer thickness and volume composition of the filament is investigated on the tensile and flexural properties of FDM printed parts. High Throughput Mechanical Analysis (HTMech) is also used to determine the tensile properties of single and bi-layer films, produced via solvent casting. The HTMech can conduct rapid tensile testing, generating a large number of data points on each sample. Results obtained will be discussed.

155a "Autonomous 3D Printing" for Novel Food and Pharmaceutical Applications

Anson W. K. Ma, Ethan Chadwick, Christopher Maiorana, Maryam Pardakhti, Shing-Yun Chang, Mingyang Tan, Pouria Hoveida, Guoan Zheng, Yushuo Niu, Qian Yang

The additive approach of 3D printing is well suited for precision medicine and personalized nutrition applications, where multiple drugs or nutrients are combined at controlled levels to tailor for individual needs. However, wider adoption of 3D printing has frequently been hampered by the relatively low throughput and the lack of validated materials. The latter further calls for more efficient material and process optimization that goes beyond a heuristics-based or grid search approach. The development costs go up quickly as the 3D printing process becomes more sophisticated with many tunable process parameters and consequently the "curse of dimensionality". In this presentation, we will first share our recent work in developing a binder jet printing (BJP) test bed for pilot-scale 3D printing of food and pharmaceutical products [Int. J. Pharm. 605, 120791 (2021)]. With the open architecture of this test bed, we have further explored a rather ambitious concept of "autonomous 3D printing" to speed up the optimization process, leveraging the latest (near) real-time, in-situ metrology and machine learning methods. [This work is partially supported by the USDA National Institute of Food and Agriculture, AFRI project 2019-06721.]

155b Achieving High Optical Translucency in SLA-Printed Elastic Materials for Patient-Specific Anatomical Models in Cardiovascular Medicine Applications

Alan Aguirre-Soto

Stereolithographic (SLA) based additive manufacturing is one of the most popular methods to create complex polymer-based models thanks to the balance between cost and overall performance (resolution, speed, broad materials library, etc.). However, there are several important challenges that researchers have generally been trying to overcome, including 1) printing with soft materials, and 2) SLA-based fabrication of models with glass-like transparency. These obstacles need to be addressed to implement this method in areas like the fabrication of patient-specific anatomical models for applications in cardiovascular medicine, such as flow visualization phantoms, medical training and pre-procedural planning. Here, we present our work in the development of a post-processing method that enables the fabrication of high-accuracy elastic (compliant) models with close to glass-like transparency. We benchmark our final models against silicon-based commercial average-anatomy models in terms of optical transparency. We also show that our final multi-materials models have mechanical properties closer to human cardiovascular tissue than PDMS, which is one of the gold standards for flow visualization techniques. Through a comprehensive characterization we demonstrate how our proprietary formulations yield surfaces with sub-1 micron surface roughness and optical transmission values of up to 70% in hollow models. Furthermore, we show how we can integrate relatively small models into 1:1 scale full body anatomical models with our formulations. In summary, we have found a method to produce anatomical models of the cardiovascular system in elastic materials with high translucency without sacrificing in resolution or fidelity and without substantially increasing the cost. We hope that this contributes to the broader implementation of the technique in hospitals and medical schools.

155d Understanding Multiphase Behavior of Additively Manufactured Lattices: Progress Towards Personalized Biomedical Tools and Platforms

Ian Woodward, Yinkui Yu, Emily Kolewe, Catherine Fromen

The extraordinary flexibility afforded by additive manufacturing has enabled a design space that caters to the needs of a variety of fields. It has seen utility in chemical engineering unit operations and process intensification, and with its potential for hierarchical structures and mass customization, it has also become an invaluable tool for developing multidisciplinary solutions to biomedical challenges. One such area of interest is pulmonary drug delivery, in which additive manufacturing can facilitate both patient-specific medical devices and *in vitro* testing platforms. While there has been increased interest in periodic 3D printed structures for fluidic applications, the field is still nascent. In particular, multiphase interactions with these structures, as well as the influence of processing steps, remain to be understood. In this work, we examine the aerosol filtration characteristics of periodic lattice structures, taking common unit cells from the literature (simple cubic and Kelvin) as a basis. With parts fabricated using the Carbon M1 platform, we observe significant differences in particle collection between lattice sizes and differential behavior within sections of a lattice column. We discuss the influence of part resolution on coating effectiveness, particle collection, and subsequent removal of particles from the lattices. Finally, we discuss the broader implications of these findings and the utility of parametric design combined with additive manufacturing in the field of pulmonary medicine. With these developments and continued work, we emphasize the potential of additive manufacturing in developing truly personalized solutions to critical biomedical challenges.

155e Masked Stereolithography Printing for Rapid Prototyping of Microfluidic Systems with Embedded Functional Components

Isteaque Ahmed, Aashish Priye

3D printing microfluidic structures for prototyping has gained a lot of attention due to several advantages such as fast production, cost-effectiveness, and versatility over traditional photolithography and PDMS mold-based microfabrication techniques. We report the application of masked stereolithography (MSLA) 3D printing approach with our multi-material UV curable resin formulation to enable single-step fabrication of intricate and functional microfluidic chips. The hybrid technique utilizes both UV curable commercially available resins and our custom resin formulations incorporating either high molecular weight Polyethylene glycol diacrylate (700 andamp; 3400) or Ethylene glycol polyether acrylate (EGPEA) as the monomer unit to achieve sub-50 micron sized positive features with tunable mechanical and optical properties. We analyze the range of achievable optical and mechanical characteristics as a function of varying monomer composition in our custom resins. We also determine optimal resin composition and printing parameters to directly print multi-level embedded rectangular microfluidic channels (eliminating the need for chip bonding steps). By incorporating multiple resins with different mechanical properties, we were able to achieve spatial control over the stiffness of the cured resin enabling us to incorporate both flexible and rigid components within a single 3D printed structure.

We demonstrate the utility of this technique by 3D printing an integrated pressure-actuated diaphragm valve (with flexible cured resin) in an otherwise rigid and clear microfluidic device that can be fabricated in a one-step process from a single CAD file. We also present a step-by-step guide and tutorial that will enable researchers to use their commercial desktop MSLA 3D printers to print functional microfluidic devices. The versatility of our fabrication technique enables one to go from an "idea stage" to a "fabricated microfluidic chip" within ~hours as opposed to traditional clean room-based microfabrication (~days), which could drastically lower the barrier to entry and greatly accelerate lab-on-a-chip research.

155f Novel Tool for Evaluating Powder Feedstock Suitability for AM Spreading Processes

Tony Thornton, Amalia Thomas, Jamie Clayton, Dan Oropeza

In Additive Manufacturing (AM) Powder Bed Fusion (PBF) processes, powders are spread into thin layers on top of each other. Between layer depositions, particles at selected points are fused or melted together to form part of the structure being built. The thickness of the subsequent layers sets the depth-resolution of the product, but also, together with the powder layer density, the thickness affects the laser or electron beam power necessary to fuse powder over the whole layer depth correctly. If the powder layer thickness and density are not reliably uniform, the final product may show defects such as porosity, lack of fusion and residual stresses, amongst others. PBF processes require a calculated balance between layer characteristics and particle fusing methods, so being able to reliably predict the characteristics of the layers formed by particular feedstock in a specific process is critical to the quality of the printed product.

Small-particle powder mixes are often used because they allow for thinner layers and therefore finer build resolutions, but they are also often related to higher inter-particle cohesive forces that hamper the formation of uniform layers. However, the performance of a fine powder in a recoating process is also influenced by particle shape and size distribution, as well as by bulk and dynamic powder behavioural properties such as flowability, density and permeability, and external factors such as recoater type and speed, entrained moisture and process temperature. It is extremely difficult to deduce a universal model for the prediction of spreading behaviour from all these variables, for all materials, and for the wide variety of process parameters used by the whole industry.

In this work we consider the specifications of a typical spreading/recoating operation (recoater type and speed, gap height, and spreading surface), and the challenges associated with assessing powders for suitability (from technology limitations to the usefulness of measurable metrics). We propose an ex-situ method for realistically replicating the spreading process, both on a flat surface, and on a previously coated powder layer, using only small sample volumes. Then, we explain a technique for quantifying the spreading performance using technology that applies an appropriate trade-off between measurement resolution and range, and realistic usefulness of results. We describe how the instrument can obtain relevant quantitative measurements proven to be effective for the assessment of powder layer uniformity and consistency, for any type of AM powder.

We present quantitative assessments of powder layer quality and feedstock suitability for several powder samples that are varied in material (two different metals and a polymer with different treatments) and particle sizes (overall range 16 – 150 μm), but all samples were produced specifically for AM processes. The bulk properties of these powders were also evaluated using an FT4 Powder Rheometer® (Freeman Technology Ltd, UK) to investigate the general relationship between flow behaviour and spreading performance metrics across such a wide variety of materials.

221a High-Resolution 3D Printing of Piezoelectric Structures Using Micro-CLIP

Siying Liu, Wenbo Wang, Luyang Liu, Xiangfan Chen

Piezoelectric materials and composites enable a wide range of energy harvesting and sensing applications owning to their intrinsic capability of converting mechanical energy to electrical energy and vice versa. Additive manufacturing (AM), also known as 3D printing, is thriving as a category of effective and robust techniques in manufacturing three-dimensionally (3D) architected piezoelectric composites comprising of inorganic/organic piezoelectric materials, and these architected structures possess the potential of delivering anisotropic piezoelectric responses upon specific structural designs. Nonetheless, the serial nature of the additive building processes results in the inherent speed-accuracy trade-off, which seriously limits the scalability and efficiency of manufacturing functional devices that require precise control of fine features. As a result, the reported progress in this field either limit themselves to macroscale devices due to relatively low printing resolution. On the contrary, micro continuous liquid interface production (μCLIP), a recently developed 3D printing technology, has been shown to create 3D geometries at very high speeds, with good surface finish, and uniform mechanical properties.

Herein, we report rapid 3D printing of architected barium titanate (BTO) composite structures with high resolution by using mCLIP. The BTO nanoparticles were surface functionalized to enable stable dispersion in poly(ethylene glycol) diacrylate (PEGDA). Resins with up to 30 wt% of functionalized-BTO (f-BTO) nanoparticles were prepared and printed continuously to yield a variety of architected structures with unprecedented longitudinal printing speeds of up to ~ 20 μm/s. The architected structures were tested with a customized piezoelectric characterization setup and demonstrate effective piezoelectric coefficient d_{33} up to 30 pC N^{-1} and piezoelectric voltage constant g_{33} up to 460 mV mN^{-1} which is comparable to other reported work. This work not only enables the rapid 3D printing of architected piezoelectric structures with tailorable properties via compositional and structural modifications, but also intrinsically enlightens its potential capability for eco- and biocompatible applications.

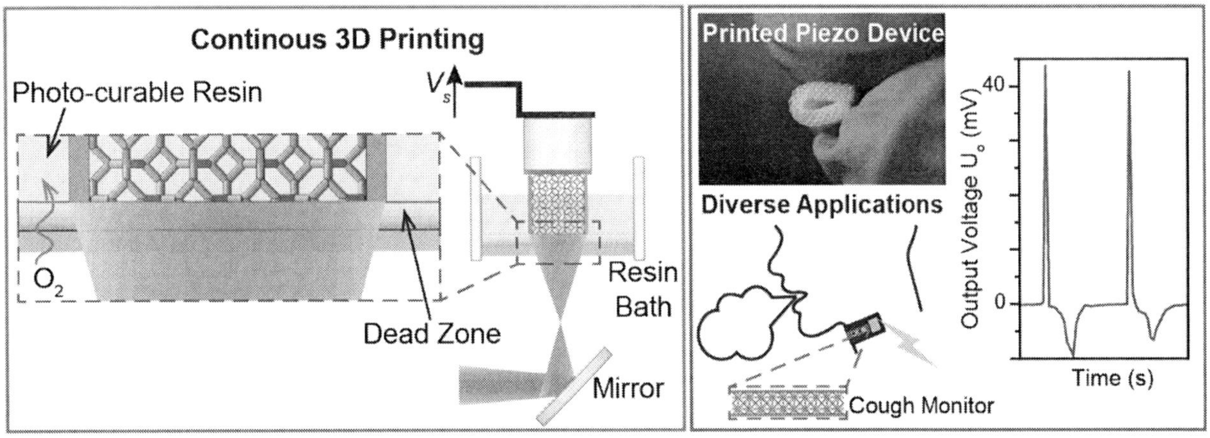

221b Enabling Off-Earth Construction and Isru through 3D Printing of Dense Regolith Suspensions

Alexandra Marnot, Blair Brettmann

The Artemis 2024 mission among other NASA-led initiatives, has demonstrated the imminent need for construction of operating bases on the Moon and later on, on Mars. While additive manufacturing (AM) methods are numerous and each provide potential avenues to enable the fabrication of necessary infrastructure, processing difficulties arising from local environments hinder the use of AM methods requiring large thermal or energy-demanding inputs. In contrast, direct ink write (DIW) AM offers a solution for construction in low-temperature, low-gravity, and low-atmospheric environments. However, the formulation factors of DIW suspensions and their effect on extrudability, solidification and shape-retention are typically not well understood when these suspensions are processed in such extreme climates. Utilizing an idealized particle system, we demonstrate a formulation suspended in a UV-curable binder that has the potential to enable AM on other planetary bodies and to optimize in-situ resource utilization. We vary the particle content, bimodal ratios, and binders, to gain insights into the printability of the model suspensions. We assess a curing formulation linked with a printing protocol that enables thorough cure of extruded layers, with solid loadings as high as 70 vol%. We test the compressive strength and particle-binder adhesion properties of parts printed and exposed to harsh thermal cycling emulating the Martian weather. By deconvoluting the characterization of these suspensions, and test-printing them under extreme environments, we show that we can expand the range of construction materials to provide cost-effective solutions for off-Earth construction.

221c 3D Printing of Thermoset/Mxene Composites with Layered Hierarchies

Sayli Jambhulkar, Kenan Song

3D printing of composites has many challenges, such as feedstock quality, printing parameter control, and structural design during layered additive manufacturing. Current 3D printing platforms include resin-based, filament-based, powder-based, and ink- or solution-based strategies. However, none of these 3D printing methods can achieve nanoscale structural controls, especially when nanoparticles are involved. Therefore, this research reports a hybrid manufacturing combining the 3D printing of stereolithography (SLA) and direct ink writing (DIW) to achieve the manipulation of nanoscale particle conformation. To demonstrate the capability of managing nanoscale features, we synthesized the MXene (i.e., $Ti_3C_2T_x$) nanoparticles via alloying metallurgy and etching dynamics. These two-dimensional flakes have a higher bending modulus than commonly used graphene and are expected to retain their 2D morphologies upon flow-assisted distribution. This distribution of nanoparticles was facilitated by (i) first the printing of surface patterns via SLA-printed topology control, and (ii) second the deposition of rheology-controlled nanoparticle suspensions. During the ink writing procedures, the secondary forces directed the anchoring and stacking of MXene with flat surfaces compliant with surface patterning. As a result, these composites display enhanced mechanical robustness, anisotropic electrical conductivity, and improved sensing capabilities, therefore useful for structural composites, microelectronics, thermal dissipators, and metasurfaces.

221d Fabrication and Evaluation of Highly Filled Polymer-Based Multi-Layer Filament for Fused Deposition Modeling (FDM)

Juhyeong Lee, Jay Park

Applying carbon-based and/or inorganic fillers in 3D printer filaments can enhance the properties of 3D-printed parts and is attracting considerable interest from academic and industry fields. Although 3D-tailored composites have been developed, very little work has been done on the production of advanced filament feedstock for FDM; particularly for highly filled 3D printer filaments that require (i) high filament strength, (ii) high dimensional accuracy, and (iii) superior surface finish. Current FDM filaments rarely exceed filler concentration of 10 wt%, for example, in case of calcium phosphate without sacrificing quality. In this work, a melt-spinning core/sheath die was designed with 2D FEM flow simulations to minimize interfacial flow instabilities; these data were generated based on the die geometry and capillary rheology of ABS and PLA with $CaCO_3$, which was then compared with the co-extruded filament. With the die, co-axial FDM filament feedstocks of amorphous or semi-crystalline polymer with up to 20 wt% filler concentration have been fabricated. Tensile bars using the filaments were successfully printed with a sufficient surface finish, respectively, and had tensile properties similar to or greater than that of neat filaments. Fracture surfaces of the printed samples after mechanical testing also elucidates the origin of the failure.

221e Direct-Write Printing of Metal-Carbon Nanotube Composites for High-Performance Electronics

Crystal Owens, A John Hart

Bulk conductors for high-performance electronics rely on metals. However, research efforts to advance the technology increasingly seek to incorporate nanomaterials such as carbon nanotubes (CNTs) due to exceptional electrical transport properties at high frequencies (>10 GHz), elevated temperatures (>200 C) and small scales (e.g., natively avoiding eddy currents through a phenomenon similar to litz wire). In this presentation, we demonstrate the development of a 3D printing process based on simultaneous electrochemical deposition of metal and dispersed CNTs within a small liquid capillary bridge between a printing nozzle and conductive substrate. This novel process can create freestanding conductors with composition and CNT alignment controlled by the applied electric field. This technique builds upon our recent development of an electrodeposition process whereby homogeneous nucleation of Cu into a CNT mesh has been used to create bulk composites in sequential processing. This has enabled control of the final conductivity of a bulk composite over several orders of magnitude up to 2 MS/m for a lightweight composite with density up to 0.8 g/cm3, and identifying the process control parameters for the nucleation of copper within CNT meshes in aqueous and organic electrolytes.

Here, we first describe the requirements to formulate a suitable printing ink, including choice of metal ions, leveling agent, base solvent and CNT functionalization to balance dispersibility in solution with final part conductivity. We further develop the process by control of the applied current in galvanostatic operation to form printed lines and 3D structures with suitable conductivity (>1 MS/m), good part homogeneity (assessed via scanning electron microscopy and electron dispersion spectroscopy) and mechanical strength (able to survive dewetting of the capillary bridge of the ink). We finally discuss the size limits to form homogeneous, conductive structures and the next steps to convert our printing method into scalable processes.

Future directions may include scalable printing of microstructured antennas and wiring onto circuit boards, and for incorporation of our CNT-laden electrochemical deposition ink into standard metal-based semiconductor electronics production.

221g Three-dimensional photochemical printing of thermally activated polymer foams

S. Eileen Seo, Younghoon Kwon, Megan T. Valentine, Craig J. Hawker

Facile, on-demand manufacturing of polymer foams with desirable properties such as mechanical strength, controlled porosity, and varied composition will enable the design of custom structural components for a variety of applications. A two-step process based on initial 3D photopolymerization, and subsequent thermal processing is reported for creating thermally activated polymer foams with tunable pore size and volume fraction. The overall resin design incorporates photopolymerizable monomers (matrix) and thermally expandable microspheres (foaming agent) to produce microporosity throughout the resulting polymer network. During the microscopic expansion process, 3D printed polymeric parts irreversibly expand while retaining their overall shape. Low-density foams with a porosity up to 85% have been produced that exhibit increased specific modulus and energy dissipation when compared to the bulk matrix and unexpanded 3D printed parts. In addition, by tuning the crosslink density of the matrix network, it is possible to control pore size and material mechanics. This approach produces thermally activated polymer foams of complex geometries with programmed porosity and enhanced mechanical properties with the two-step process also providing local control over the spatial distribution of pore sizes through multimaterial 3D printing.

285a Using Additive Manufacturing to Advance the Design of Water Treatment Devices

William Phillip

The use of non-traditional sources of water (e.g., seawater and wastewater) is essential to supporting human life on earth. While seawater desalination will remain an indispensable technology, its large energy demand necessitates the development of complementary wastewater reclamation and reuse strategies. Fit-for-purpose water, one envisioned opportunity, utilizes distributed treatment systems to provide water for localized reuse at the potable or non-potable quality demanded by its intended application. In this manner, the separation is tailored based on the chemical composition of the source water with the end-use requirements in mind. Separation devices fabricated from block polymers provide a versatile platform that can help address the technical hurdles associated with this emerging paradigm. In particular, the performance of the block polymer membranes can be controlled through thoughtful design of the macromolecular precursors to effect selective separations through affinity-based adsorption, electrostatic interactions, and catalytic conversion of target solutes. In this talk, we will discuss how additive manufacturing can be used 1) to transform block polymer materials into high-affinity, high-capacity sorbents with hierarchical structures and 2) to access emergent transport mechanisms through charge-patterning of block polymer membranes. Through these examples, we will demonstrate that additive manufacturing provides several opportunities for re-envisioning the design and fabrication of multifunctional separation devices that are molecularly engineered for the purification and conservation of water resources.

285b 4D Printing of Hydrogels Displaying Swelling-Induced Surface Wrinkling Patterns

Murat Guvendiren

Additive manufacturing of stimuli responsive materials leads to 3D printed constructs that can adapt their properties in response to an external stimuli. This process is referred as 4D printing. 4D printing is commonly used to fabricate shape-shifting constructs, which change their shape after printing into a preprogrammed shape. Although 4D printing of shape-shifting objects have significantly expand the applications of 3D printing, novel ink formulations are needed for 4D printing to create 2D and 3D structures that allow dynamic control of properties beyond shape change. In this study, we present smart inks for material extrusion-based direct ink writing (DIW) printing that display preprogrammed surface wrinkling patterns on 2D and 3D surfaces post-printing, which appear only when exposed to a solvent. Hydrogels displaying surface wrinkling patterns with user-defined order and shape are useful for a range of applications, including microfluidics, flexible electronics, smart adhesives, optical sensors, and cell culture platforms. However, conventional techniques to fabricate wrinkling patterns involve multiple steps and are significantly limited to create hierarchical patterns, wrinkles on 3D and non-planar structure. It is also very difficult to scale-up the manufacturing process and to integration it into a continuous manufacturing process. 4D printing is used in this work to overcome these issues. The use of photocurable ink formulations allow us to tailor flow properties and printability by controlling photopolymerization time and partial crosslinking time. The crosslinker concentration is varied to develop distinct ink formulations with preprogrammed wrinkling morphology, including lamellar, peanut and hexagonal patterns. We demonstrate 4D printing of optical devices, solvent sensors, anti-counterfeit devices and high-throughput cell culture platforms.

285c Multiphoton Lithography of Organic Semiconductor Devices for 3D Printing of Flexible Electronic Circuits, Biosensors, and Bioelectronics

Mohammad Reza Abidian

In recent years, 3D printing of electronics have received growing attention due to their potential applications in emerging fields such as nanoelectronics and nanophotonics. Multiphoton lithography (MPL) is considered the state-of-the-art amongst the microfabrication techniques with true 3D fabrication capability owing to its excellent level of spatial and temporal control. Here we introduce a homogenous and transparent photosensitive resin doped with an organic semiconductor material (OS) compatible with MPL process to fabricate variety of 3D OS composite microstructures (OSCMs) and microelectronic devices. Inclusion of 0.5 wt% OS in the resin enhanced the electrical conductivity of the composite polymer about 10 orders of magnitude and compared to other MPL-based methods, the resultant OSCMs offered high specific electrical conductivity. As a model protein, laminin was incorporated into these OSCMs without a significant loss of activity. The OSCMs were biocompatible and supported cell adhesion and growth. Glucose oxidase encapsulated OSCMs offered a highly sensitive glucose sensing platform with nearly 10-fold higher sensitivity compared to previous glucose biosensors. In addition, this biosensor exhibited excellent specificity and high reproducibility. Overall, these results demonstrate the great potential of these novel MPL-fabricated OSCM devices for a range of applications from flexible bioelectronics/biosensors, to nanoelectronics and organ-on-a-chip devices.

285d 3D Synthetic Brain: Manufacturing, Challenges, and Fundamental Studies

Lawrence Ray, Caleb Shaw, Terence C. Burns, Dr. M. Rashed Khan

3D printing-based additive manufacturing technologies have been used to produce 3D synthetic brains from thermoplastic polymers that are primarily rigid. Bioprinting technologies are continuously advancing the current knowledge for in-vitro studies utilizing hydrogel-based biomaterials; yet, manufacturing complex three-dimensional (3D) shapes (e.g., miniaturized 3D brain) from soft and squishy hydrogels remains a fundamental challenge. While current studies utilizing rigid 3D brains aim for teaching, surgical planning, and training protocols, we demonstrate a subject-specific miniaturized synthetic 3D brain manufacturing for proof-of-concept studies utilizing soft and squishy biomaterial, i.e., Agarose-gel. 0.6% (w/v) and 2% (w/v) Agarose-gel shares tissue-specific mechanistic details of healthy and diseased brains that we targeted to demonstrate the utility of the manufactured 3D synthetic brain. Manufacturing 3D brains harnessing pure Agarose-gels enable molecular diffusion studies, which previously remained a gap in the literature that we overcome. Towards that goal, subject-specific magnetic resonance imaging (MRI) data were evaluated, segmented, and reconstructed to produce a volumetric STL file; then, we produced an elastomeric inverse replica mold from a rigid 3D printed brain utilizing the volumetric STL data. Pouring and gelling Agarose-gels within the inverse replica mold produce the soft and squishy brains onto which Methylene Blue was infused to investigate time-dependent molecular diffusion. Beyond molecular diffusion, a soft and squishy 3D brain from Agarose can provide platforms for other studies relevant to drug delivery, neural stimulation, neuroelectronics, and neurofluidics.

285e Additive Manufacturing of MOF Contactors for CO_2 Capture

Hannah Holmes, Wenying Quan, Simon C. Weston, Carter W. Abney, William J. Koros, Matthew Realff, Ryan P. Lively

Metal-organic frameworks (MOFs) with high CO_2 capacities have been developed for CO_2 capture technologies, but their full performance is often not realized in pellet-packed beds due to high pressure drops, slow mass transfer, lack of thermal management (i.e., cooling during adsorption), and inefficient heating and cooling during the desorption and cooling steps respectively. Implementing MOFs in structured adsorbent contactors with thermal management and heat integration (e.g., sensible heat recovery post-desorption) mitigates the limitations.

Here we utilize solution-based additive manufacturing (SBAM), a form of 3D printing, to directly fabricate hierarchically-porous contactors containing a MOF. Contactor porosity is achieved via solvent-evaporation induced phase inversion of a ternary polymeric solution during the printing process. The dependence of porosity on printing parameters, including nozzle temperature, bed temperature, and printing speed, is demonstrated. Functionalization of the MOF is performed post-printing, and the geometrical versatility of SBAM is demonstrated through novel contactor designs. Pre- and post-printing characterization of the MOF, including CO_2 isotherms, N_2 physisorption, XRD, NMR, and SEM, confirm its stability throughout the fabrication process. The contactor performance is evaluated for natural gas combined-cycle (NGCC) applications using dynamic humid CO_2 breakthrough experiments. Overall, the results of this work provide a foundation for fabricating MOF contactors with enhanced mass and heat transfer for humid CO_2 capture applications.

285f Modeling of Sagging for 3D Printed Layers during the Curing Process

Andrey Filippov, Jeremy M. Lenhardt, Todd H. Weisgraber, Thomas S. Wilson, Fangyou Xie, Lemuel X. Perez Perez, Andrew L. Nguyen, Steven J. Guzorek, Hamed Z. Ammar

Modeling of Sagging for 3D Printed Layers During the Curing Process.

Andrey Filippov (Presenting Author), Jeremy Lenhardt, Todd Weisgraber, Thomas Wilson, Fangyou Xie, Lemuel Perez Perez, Andrew Nguyen, Steven Guzorek, and Ziad Ammar

Lawrence Livermore National Laboratory, 7000 East Ave, Livermore, CA 94550

Extrusion-based 3D printing offers a quick way to generate 3D polymer structures without the need for molds. Through precise control over the deposition of highly engineered viscoelastic inks in the form of continuous filaments, complex 3D architectures can be additively fabricated, layer-by-layer, to generate parts that are directly applicable to cutting-edge technologies. 3D printed silicones are well suited to soft materials applications including high-performance foams and cellular solids (Maiti *et al.*, 2016) soft robots (Mac Murray *et al.*, 2017) biomedical devices and biological scaffolds (Barry *et al.*, 2009) and flexible and stretchable electronics (Rim *et al.*, 2016)

DIW offers the advantage of producing fine composition gradations since the nozzle can accept more than one ink and the relative flow rate of each ink can be independently and arbitrarily adjusted in time by means of a computer-controlled dispensing system.

Careful control over the chemistry, network formation, and crosslink density of the ink formulations can be used to tune the mechanical properties of 3D printable silicones and improve their stability. (Durban *et al.*, 2017). Nonpolar silicone materials are not naturally pseudoplastic or thixotropic; however, a hydrophilic solid filler such as untreated fumed silica can be incorporated to render a solid-like network in the absence of applied stress, reinforcing the matrix. The final stage of the DIW process is the curing, which is usually induced either by UV irradiation or thermosetting.

Thermosetting is a popular method of fixing 3D-printed structures, 'freezing' them in desired shapes by gradual heating in the ovens. Unfortunately, during the time between the structure printing and the curing, the shape of 3D structures can change (sag) due to effects of gravity, which is more common for larger structures, and the surface tension, typical for smaller scale structures. Evaluation and detailed prediction of printed structure deformations is a challenging problem, requiring creation of complex representative mathematical and numerical models and their efficient solutions.

These models should account for non-Newtonian properties of the printed material, predict dynamics of the varying shape of the interface between the polymer and the air, and transient heat transfer and spatially non-uniform polymerization in complex 2-phase structures. Thermosetting of the polymer results in dramatic effect of its effective viscosity, practically stopping the plastic deformation of the lattice. All these phenomena are essentially coupled, which results in additional numerical difficulties as the corresponding time scales are usually very different.

In present study, the printed layer of silica-enforced PDMS (polydimethylsiloxane-co-diphenylsiloxane) is modeled as a two-phase system consisting of air and the polymer with interface set up during the printing process. The structure geometry, which initially one of several lattice configurations, changes mostly due to the action of the surface tension, while all material properties are strongly temperature dependent. The polymer flow is described by equations of the extended Herschel-Bulkley model with parameters strongly dependent on temperature and degree of curing.

To determine parameters of the model, several experiments have been carried out. The parameters of the extended Herschel-Bulkley model were determined using flow sweep measurements and separate experiments with vertical structure sagging at different temperatures. To model the curing process, the dependences between the curing rate, degree of curing and temperature were obtained in a dedicated experiment using differential scanning calorimetry (Hong and Lee, 2012).

Fig.1 An example of simulated initial (a) and deformed lattice (b) (Attached)

Based on the mathematical model, the numerical model was build using commercial package ANSYS-FLUENT with material properties set up by corresponding user-defined functions. The model is used for simulations of printed structure sagging with different initial and boundary conditions and compared with experimentally observed lattice deformations. For example, Fig. 1 shows a simulated sagging deformation of a 3d printed lattice.

References

Barry, R.A., Shepherd, R.F., Hanson, J.N., Nuzzo, R.G., Wiltzius, P. and Lewis, J.A. (2009) Direct-write assembly of 3D hydrogel scaffolds for guided cell growth. Adv. Mater. **21**, 2407–2410.

Mac Murray, B.C., An, X., Robinson, S.S., van Meerbeek, I.M., O'Brien, K.W., Zhao, H. and Shepherd R.F. (2015) Poroelastic foams for simple fabrication of complex soft robots. Adv. Mater **27**, 6334.

Maiti, A., Small, W., Lewicki, J.P., Weisgraber, T.H., Duoss, E.B., Chinn, C., Pearson, M.A., Spadaccini, C.M., Maxwell, R.S. and Wilson, T.C. (2016) 3D printed cellular solid outperforms traditional stochastic foam in long-term mechanical response. Sci Rep. **6**, 24871.

Durban, M.M., Lenhardt, J.M., Wu, A. S., Small, W.IV, Bryson, T.M., Perez-Perez, L., Nguyen, D.T., Gammon, S., Smay, J.E., Duoss, E.B., Lewicki, J.P. and Wilson, T.S. (2017) Custom 3D printable silicones with tunable stiffness. Macromol. Rapid Commun. 1700563

Rim, Y.S., Bae, S.H., Chen, H.J., De Marco, N. and Yang, Y. (2016) Recent progress in materials and devices toward printable and flexible sensors. Adv.Mater.2016, 28, 4415–4440.

Hong, I. and Lee, S. (2013) Cure kinetics and modeling the reaction of silicone rubber. Jour. Ind. Eng. Chemistry **19**, 42-47.

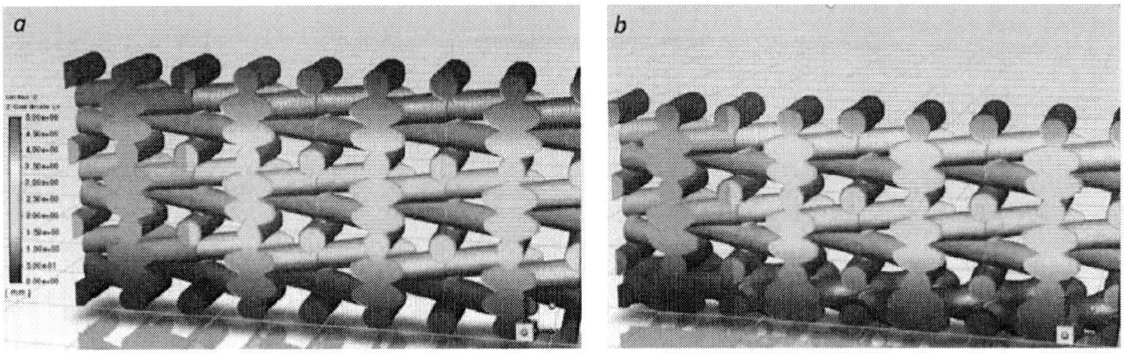

Fig.1 An example of simulated initial (a) and deformed lattice (b)

472a Keynote Talk-Self-Driving Fluidic Micro-Processors for Accelerated Discovery and Manufacturing of Energy Materials

Milad Abolhasani

Despite the intriguing properties and widespread applications of colloidal semiconductor nanomaterials in energy technologies, their discovery, synthesis, and manufacturing are still based on Edisonian trial-and-error based techniques. Existing materials development strategies using resource-intensive batch reactors with irreproducible and uncontrollable heat/mass transport rates very often fail to overcome the demands of the vast synthesis and processing universe of energy materials, resulting in a slow and expensive discovery and development timeframe (8-10 years). Recent advances in flow chemistry and machine learning (ML)-guided modeling/decision-making strategies provide an exciting opportunity to reshape the discovery and manufacturing of emerging solution-processed energy materials.[1] In this talk, I will present an end-to-end 'self-driving lab' for autonomous discovery, development, and on-demand manufacturing of energy materials through convergence of flow chemistry, robotics, and *in-situ* material characterization with ML.[2] I will discuss how modularization of different stages of materials synthesis and processing in tandem with an ensemble neural network modeling and decision-making under uncertainty can enable a resource-efficient navigation through a high dimensional experimental design space of clean energy nanomaterials. An application of the self-driving lab for the autonomous synthesis of metal halide perovskite quantum dots will be presented.

References

1. (a) Volk, A. A.; Abolhasani, M., *Trends in Chemistry* **2021**, *3*(7), 519-522.; (b) Volk, A. A., et al., *Advanced Materials* **2020,** *33* (4), 2004495.
2. (a) Abdel-Latif, K., et al., *Advanced Intelligent Systems* **2021,** *3* (2), 2000245; (b) Epps, R. W., et al., *Chemical Science* **2021,** *12* (17), 6025-6036.

472b Keynote Talk-Advancements in Manufacturing Via Strategic Assemblies of Bio-Based Polymers and Composites and Adaptive 3D Printing Techniques

Joseph Stanzione III

The utilization of wood-derived building blocks and other abundantly available, non-food biomasses to replace fossil-based precursors is an attractive research subject in modern polymer science and engineering. The continuing development of economically viable bio-based feedstocks and subsequent, new platform chemicals opens opportunities to increase the resource base from which chemical and materials engineers can draw to meet specific polymer and composite property requirements. Recently, significant efforts have transpired to obtain bio-based polymers and composites with enhanced thermomechanical properties for a wide variety of applications, from adhesives to coatings to under-the-hood, lightweight, yet geometrically complex vehicle components. In continuing these efforts, we have prepared and examined a variety of bio-based thermoplastics, thermosets, and composites, including epoxies, vinyl esters, (meth)acrylates, allyl ethers, polyesters, and polycarbonates with and without reinforcing fibers. These novel materials were strategically assembled from monomers based on lignin-derived, (hemi)cellulose-derived, soy-derived, and cashew nut shell liquid-derived platform chemicals, among others. Furthermore, these novel materials were processed using adaptive 3D printing techniques, including vat photopolymerization, direct write, and cold spray additive manufacturing techniques. This talk will highlight our work on fundamentally understanding the processing-structure-property relationships of these bio-based polymers and composites via thermal, mechanical, and modelling techniques. Additionally, their utilities in military and civilian applications will be discussed along with an identification of critical technological gaps that should be investigated through advanced modelling and experimental techniques.

472c Keynote Talk-Advanced Membrane Manufacturing for Water and Energy Applications

Oishi Sanyal

Membrane-based processes could offer significant energy savings for large-scale separations as has been demonstrated in case of desalination where reverse osmosis has largely replaced thermal processes. For applications beyond desalination, appropriate manufacturing strategies must be pursued in addition to advanced material development to broaden the application scope of membranes. In this talk, I will describe two membrane manufacturing techniques – hollow fiber spinning and direct ink-writing and highlight their advantages and challenges. The 1st part of my talk will describe the formation of polyimide-derived carbon molecular sieve (CMS) hollow fiber membranes and elaborate on the fundamental challenges in creating high performance CMS membranes for gas separation. The 2nd half of my talk will focus on utilizing direct ink-writing to develop porous microfiltration/ultrafiltration membranes. This new technology provides opportunities to reduce the solvent usage associated with conventional fabrication methods, while enabling precise control on the membrane morphology.

472d A Physics-Informed Machine Learning Model for Battery Capacity Fading Prediction with Early Cycling Data

Jiwei Yao, Qiang Gao, Benben Jiang, Kody Powell, Tao Gao

High-performance lithium-ion battery (LIB) technology is a key component to achieving high penetrations of renewable energy. However, there are numerous challenges that still need to be solved, such as increasing the energy density from 250 Wh/kg to 500 Wh/kg, achieving fast charging, and extending the cycle life to above 10,000 cycles. Much research has been carried out on addressing material problems, developing advanced battery management systems, and designing modern cells [1–4]. However, relatively little efforts have been made on developing battery models, especially that can predict capacity fading. Given the current capacity fading models, we can classify them into two categories: electrochemical models and regression model [5–7]. Despite the incorporation of electrochemical principles and newfound understanding of capacity fading mechanisms, the well-known electrochemical model, the pseudo-2D model, still cannot provide accurate prediction of capacity fading. For the regression model, thanks to the recent advancements in machine learning techniques, many models have been proposed. However, many of these models are physics-blind (i.e., pure black box), data-driven models which are unable to adapt to the changing internal environment.

To address these challenges, in this study, we proposed a physics-motivated, data-driven model to predict capacity fading during battery cycling. The pipeline is shown in Figure 1. *For Li-ion batteries, there are two major contributors to the loss of active Li-ion inventory: the growth of a solid-electrolyte interface (SEI) and Li plating. Depending on which fading mechanism is dominant, the battery will experience a different fading rate. Compared with the growth of SEI, Li plating will result in rapid capacity fading [8]. To distinguish the dominant fading mechanism within the battery,* based on the electrochemical principles of the Li-ion battery, we propose using, ΔV, *as the primary feature for fading classification. The K-means clustering technique is employed and draws a boundary at 15.6 mV. With their average capacity fading at the* 50th *cycle, batteries with* $\Delta V < 15.6\ mV$ *are denoted as the fast-fading group, others are denoted as the slow-fading group.*

For the slow-fading group, the loss of capacity is caused by the growth of SEI. Therefore, the resistance of the SEI is a good feature for capacity fading prediction. However, there is no way to attain the resistance of the SEI from the cycling data. Instead of obtaining resistance of the SEI from the cycling data, which is not feasible, in this study, we propose using the internal resistance (R) as a substitute feature obtained from current-voltage curve. In contrast, Li plating is the dominant contributor to the capacity loss for batteries in the fast-fading group. Therefore, ΔV is chosen as the regression feature. To validate the model performance, a benchmark with log variance of the discharging curve difference $\Delta Q_{N-1}(V)$ is selected [9].

Given data from the 5th cycle, we achieve 19% and 26% test mean percentage error (MPE) for batteries in the slow-fading group and in the fast-fading group, while the benchmark has 334% test MPE. Furthermore, by adding the prospective cumulative efficiency and past average fading rate as the second and third feature, we achieve 7%, 11%, and 40% test MPE for slow fading, fast fading, and the benchmark, respectively.

Reference

[1] Sheha MN, Powell KM. Dynamic Real-Time Optimization of Air-Conditioning Systems in Residential Houses with a Battery Energy Storage under Different Electricity Pricing Structures. vol. 44. Elsevier Masson SAS; 2018. https://doi.org/10.1016/B978-0-444-64241-7.50416-X.

[2] Sheha MN, Powell KM. An economic and policy case for proactive home energy management systems with photovoltaics and batteries. Electr J 2019;32:6–12. https://doi.org/10.1016/j.tej.2019.01.009.

[3] Hesse HC, Martins R, Musilek P, Naumann M, Truong CN, Jossen A. Economic optimization of

component sizing for residential battery storage systems. Energies 2017;10. https://doi.org/10.3390/en10070835.

[4] Suo L, Borodin O, Sun W, Fan X, Yang C, Wang F, et al. Advanced high-voltage aqueous lithium-ion battery enabled by "water-in-bisalt" electrolyte. Angew Chemie 2016;128:7252–7.

[5] Torchio M, Magni L, Gopaluni RB, Braatz RD, Raimondo DM. LIONSIMBA: A Matlab Framework Based on a Finite Volume Model Suitable for Li-Ion Battery Design, Simulation, and Control. J Electrochem Soc 2016;163:A1192–205. https://doi.org/10.1149/2.0291607jes.

[6] Yao J, You F. Simulation-based optimization framework for economic operations of autonomous electric taxicab considering battery aging. Appl Energy 2020;279:115721. https://doi.org/10.1016/j.apenergy.2020.115721.

[7] Yang XG, Leng Y, Zhang G, Ge S, Wang CY. Modeling of lithium plating induced aging of lithium-ion batteries: Transition from linear to nonlinear aging. J Power Sources 2017;360:28–40. https://doi.org/10.1016/j.jpowsour.2017.05.110.

[8] Gao T, Han Y, Fraggedakis D, Das S, Zhou T, Yeh CN, et al. Interplay of Lithium Intercalation and Plating on a Single Graphite Particle. Joule 2021;5:393–414. https://doi.org/10.1016/j.joule.2020.12.020.

[9] Severson KA, Attia PM, Jin N, Perkins N, Jiang B, Yang Z, et al. Data-driven prediction of battery cycle life before capacity degradation. vol. 4. 2019. https://doi.org/10.1038/s41560-019-0356-8.

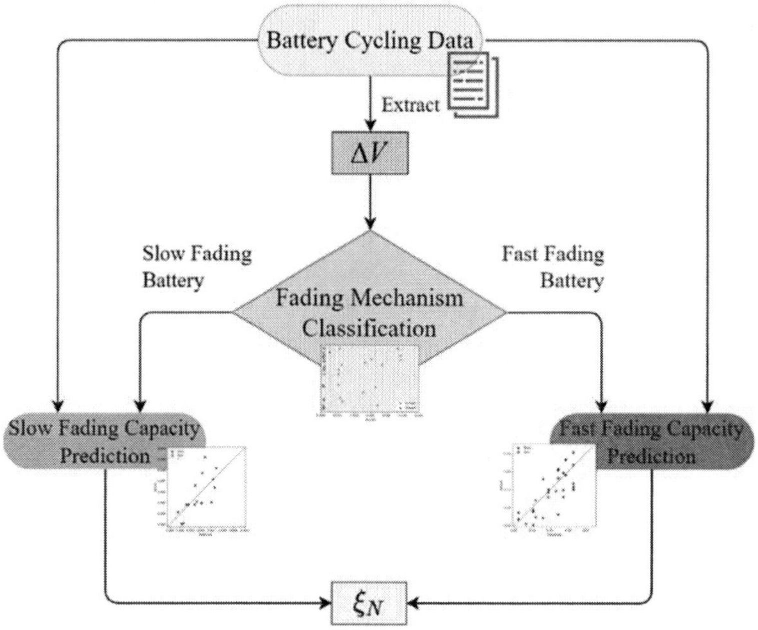

Figure 1. Pipeline of the physics-motivated data-driven model, comprising three modules, fading mechanism classification module, fading prediction module for slow fading group, and fading prediction module for fast fading group

472e On Testing Methods for Image-Based Control Systems for Next-Generation Manufacturing

Henrique Oyama, Fnu Akkarakaran Francis Leonard, Katie Tyrrell, Helen Durand

Next-generation manufacturing utilizes process automation and data to improve system performance. This can include new data modalities such as image-based sensors with image processing algorithms integrated with typical process automation concepts, which can enable controllers based on camera sensors to measure process variables that would otherwise be impractical to be collected online (e.g., cell counting which is traditionally measured offline). These image-based control (IBC) systems have been used in applications that range from bioprocess monitoring [1] and drug discovery [2] to robot manipulators [3] and autonomous driving systems [4]. Although significant advances in visual feedback control using image-based sensors and image processing algorithms have been reported in the literature [5][6][7], these control architectures integrated with a vision system may be potential targets of cyberattacks that aim to falsify image-based measurements provided to the controller. In light of this, simulation-based studies involving image data modality are needed to understand the impacts of cyberattacks on image-based control systems and propose control/detection strategies to detect and prevent malicious actions.

To enhance the performance of autonomous systems, different IBC systems have been proposed using a variety of monitoring systems and control architectures, including classical proportional-integral-derivative (PID) controllers [3] and advanced control formulations (e.g., model predictive control (MPC) [6]). In [3], an hexarotor image-based visual servo control has been proposed using a neural network-based PID controller integrated with a visual system, where physical data was collected and simulations were performed in MATLAB via the Robotics Toolbox [8]. In [6], an image-based control design using a nonlinear MPC has been developed as a strategy for an unmanned aerial vehicle, where the constraints related to actuator and visibility limitations can be added to the MPC formulation. For applications in the chemical process industry, [9] describes a process monitoring system for an industrial boiler based on a real-time image analysis framework. In [10], a real-time image analysis system has been used to estimate bubble sizes at a phosphorus oxide flotation process. In [11], bubbles in fluids have been simulated as a 3D animation with physical realism by considering the physics of dissolved gas diffusion, nucleation, and their interaction with liquids.

Moving towards realistic simulations of IBC systems, 3D programming software such as Blender, which has an embedded Python interpreter, may help with creating and testing IBC systems for different industrial applications before being implemented on the real system, which can offer a manner to test how new next-generation manufacturing strategies such as image-based control schemes might behave in the actual process. Motivated by the advances in visual feedback control systems and animation, we showcase how Blender can be used to develop and test IBC designs under different control strategies, and evaluate the implications of different cyberattack events involving image data tampering on process safety. Simulations of level control in a tank are used to illustrate the proposed methodology. The level in the tank is simulated in the Blender Python programming interface, and this information is used to adjust the position of the height of the tank (modeled as the top of a plane) in Blender. Then, a render of the image is taken and subsequently processed via Pillow [12] to enable the pixel corresponding to the top of the tank to be identified. This is then used to estimate the height of the tank, which is fed to a proportional-integral (PI) controller to control the process. The process is simulated in the absence of a cyberattack, and also the presence of attacks, to demonstrate how Blender can be used to provide evaluations of image-based control laws under different conditions. Finally, we discuss how Blender can also be used to add noise to an image to give it a characteristic that might be a part of other cyberattack detection algorithms [13].

References:

1. Bluma, Arne, Tim Höpfner, Patrick Lindner, Christoph Rehbok, Sascha Beutel, Daniel Riechers, Bernd Hitzmann, and Thomas Scheper. 'In-situ imaging sensors for bioprocess monitoring: state of the art.' Analytical and Bioanalytical Chemistry 398.6 (2010): 2429-2438.
2. Chandrasekaran, Srinivas Niranj, Hugo Ceulemans, Justin D. Boyd, and Anne E. Carpenter. 'Image-based profiling for drug discovery: due for a machine-learning upgrade?.' Nature Reviews Drug Discovery 20.2 (2021): 145-159.
3. Su, Yuxin, and Chunhong Zheng. 'A simple PID control for asymptotic visual regulation of robot manipulators.' International Journal of Robust and Nonlinear Control 21.13 (2011): 1525-1540.
4. Bengler, Klaus, Klaus Dietmayer, Berthold Färber, Markus Maurer, Christoph Stiller, and Hermann Winner. 'Three decades of driver assistance systems: Review and future perspectives.' IEEE Intelligent Transportation Systems Magazine 6.4 (2014): 6-22.
5. Lopez-Franco, Carlos, Javier Gomez-Avila, Alma Y. Alanis, Nancy Arana-Daniel, and Carlos Villaseñor. 'Visual servoing for an autonomous hexarotor using a neural network based PID controller.' Sensors 17.8 (2017): 1865.
6. Lee, Daewon, Hyon Lim, and H. Jin Kim. 'Obstacle avoidance using image-based visual servoing integrated with nonlinear model predictive control.' 2011 50th IEEE Conference on Decision and Control and European Control Conference. IEEE, 2011.
7. Hajiloo, Amir, Mohammad Keshmiri, Wen-Fang Xie, and Ting-Ting Wang. 'Robust online model predictive control for a constrained image-based visual servoing.' IEEE Transactions on Industrial Electronics 63.4 (2015): 2242-2250.
8. Corke, Peter I. 'A robotics toolbox for MATLAB.' IEEE Robotics andamp; Automation Magazine 3.1 (1996): 24-32.
9. Yu, Honglu, and John F. MacGregor. 'Monitoring flames in an industrial boiler using multivariate image analysis.' AIChE Journal 50.7 (2004): 1474-1483.
10. Lin, Bao, Bodil Recke, Jørgen K. H. Knudsen, and Sten Bay Jørgensen. 'Bubble size estimation for flotation processes.' Minerals Engineering 21.7 (2008): 539-548.
11. Cleary, Paul W., Soon Hyoung Pyo, Mahesh Prakash, and Bon Ki Koo, 'Bubbling and frothing liquids.' ACM SIGGRAPH 2007 papers. 2007. 97-es.

12.Clark, A. (2015). *Pillow (PIL Fork) Documentation.* readthedocs. Retrieved from https://buildmedia.readthedocs.org/media/pdf/pillow/latest/pillow.pdf

13. H. Farid, "Digital image forensics: lecture notes, exercise, and matlab code for a survey course in digital image and video forensics," p. 199, 2012, accessed: 2021-12-10.

472f Comparative Study for Evaluation of Thermal Integration Projects Based on Pinch Technology

Ícaro Almeida, Beatriz Dantas, Fernanda Andrade, Heloyse Reges, Wilton Lima, Fernando V. Lima, Heleno Bispo

Thermal integration in chemical processes has proved to be an effective alternative, given the continuous pressure imposed by regulatory standards. Additionally, it is well-known that the consumption of thermal energy in the industry contributes to a significant fraction of CO2 emissions and other greenhouse gasses. Moreover, the correlation between the rising energy prices and potential carbon/energy taxes invariably leads to higher operating costs[1]. However, process stream calorific potentials can be applied to reduce the use of utilities and fuels, resulting in lower emissions, residues, and operational costs.

Pinch analysis has proven to be a successful solution for reducing energy consumption in numerous industries worldwide. The knowledge about this technique has expanded far beyond the original studies and has become a consolidated part of the global strategy for process design optimization and associated advances[2]. By evaluating the hot and cold process streams, a deliberative trade-off between operating and capital costs is taken into account to obtain the minimum total cost in a heat-exchanger network (HEN)[3]. Such an approach allows the definition of requirements for heating and cooling based on a deep understanding of the process from an energy perspective.

Although some commercial process simulators automate the calculation procedures related to Pinch technology (such as HYSIM®, HEXTRAN®, and Aspen Energy Analyzer®), these tools demand a high acquisition cost, being a possible limiting factor for their disseminated use by the scientific and engineering communities. Motivated by this, this research evaluates the potential of the Heat-Exchanger Network Synthesis Analysis and Design (HENSAD) software as an open-access tool based on Pinch analysis for establishing energy and economic targets, in addition to the cost estimation for heat exchanger networks.

For this purpose, the software infrastructure is established upon the data provided by a real-time acquisition system (PI System - AVEVA/OSIsoft) through an integrated platform with process simulators (Aspen Plus, Unisim, or DynSim) and pre-treatment of thermodynamic data to quantify the heat potential of process stream fractions developed in Python. A comparative study is performed to validate the results generated by the Aspen Energy Analyzer (AspenTech) and HENSAD. Such evaluations show the effectiveness of HENSAD to provide valuable economic analysis.

References

[1] Olsen, D., Abdelouadoud, Y., Liem, P., Welling, B. The Role of Pinch Analysis for Industrial ORC Integration. Energy Procedia, v. 129, p. 13-15, 2017.

[2] Kemp, I. C. Pinch Analysis and Process Integration: A user Guide on Process Integration for the Efficient Use of Energy. 2. Ed. Elsevier, 2007.

[3] Turton, R., Bailie, R. C., Whiting W. B., Shaeiwitz J. A. Analysis, Synthesis, and Design of Chemical Processes. 4. Ed. Prentice Hall, 2012.

472g Dynamic Characteristics Investigation and Control of Tubular Reactor with Supercritical Water Gasification

Cui Wang, Hui Jin, Zhe Wu

Hydrogen energy is an important source of energy that cannot be obtained directly from nature [1]. Supercritical water gasification (SCWG) technology is a potential clean and efficient technology to convert biomass waste to hydrogen rich gas, which has received increasing attention in recent years, for example, the research works on gasification performance of different feedstocks, gasification mechanism, catalyst study and reactor design [2, 3, 4]. With regards to the reactor design, dynamic performance of continuous reactors such as tubular reactor and fluidized bed reactor is a research area that is of great importance for the implementation of this technology in industry. The existing works on reactor dynamic performance mainly focused on normal operating condition [5]; however, the dynamic characteristic of supercritical water gasification reactor could be different from the reactor under normal operating conditions due to unique properties of supercritical water.

To address the above issue, supercritical water gasification experiment of biomass will be first conducted in this work to obtain gasification mechanism and detailed reaction pathway. Then the kinetic model of biomass gasification in supercritical water will be developed based on the experimental results accounting for the reaction pathway. To evaluate the performance of reactor and kinetic model, kinetic data in literature will be adopted in this reactor, and the obtained results will be compared with simulation results. Additionally, sensitivity analysis of reactor will be carried out to identity the parameters that greatly influence the gasification results. Dynamic characteristic of reactor such as the response of key output parameters including hydrogen mole fraction and temperature to the perturbation of various parameters such as flowrate will be analyzed based on the sensitivity analysis. Finally, a data-driven based model predictive control system will be developed to improve the dynamic performance of the reactor.

References:

[1] P.P. Edwards, V. Kuznetsov, W.I. David. Hydrogen energy. Philosophical Transactions of the Royal Society A: Mathematical, Physical and Engineering Sciences. 365 (2007) 1043-56.

[2] K. Babaei, A. Bozorg, A. Tavasoli. Hydrogen-rich gas production through supercritical water gasification of chicken manure over activated carbon/ceria-based nickel catalysts. Journal of Analytical and Applied Pyrolysis. 159 (2021) 105318.

[3] C.R. Correa, A. Kruse. Supercritical water gasification of biomass for hydrogen production Review. Journal of Supercritical Fluids. 133 (2018) 573-90.

[4] J. Chen, Y. Fan, X. Zhao, E. Jiaqiang, W. Xu, F. Zhang, et al. Experimental investigation on gasification characteristic of food waste using supercritical water for combustible gas production: Exploring the way to complete gasification. Fuel. 263 (2020) 116735.

[5] H.-H. Lee, J.-C. Lee, Y.-J. Joo, M. Oh, C.-H. Lee. Dynamic modeling of Shell entrained flow gasifier in an integrated gasification combined cycle process. Applied energy. 131 (2014) 425-40.

419a Keynote Talk-on the Role of Control System Design on Detecting Cyberattacks

Matthew Ellis, Shilpa Narasimhan, Nael El-Farra

The increasing reliance on advanced communications and networking technologies and the increasing complexity and sophistication of cyberattacks have rendered industrial control systems, such as process control systems, vulnerable to cyberattacks by malicious agents [1-2]. Cyberattacks on control systems aim to disrupt operations, destabilize processes, or cause safety incidents [1-3]. A cyberattack against a control system may manipulate or disrupt the data communicated over the links between the sensors and controller or the controller and actuators. These attacks compromise the control system integrity, causing it to operate based on falsified or inaccurate data. The traditional approach for addressing the threat of cyberattacks involves fortifying information technology (IT) systems to prevent attackers from gaining access to the control system network. As evident by the myriad of recent cyberattacks, IT-based approaches are insufficient in preventing cyberattacks, and operational technology-based approaches are needed.

Detecting the presence of a cyberattack targeting the control system is one necessary OT-based capability. Cyberattack detection methods detect the presence of an attack by identifying abnormal operating behavior from operational data [4-6]. Several methods have been proposed, including chi-squared, cumulative sum, and data-driven and machine learning-based approaches (see, for example, the reviews [4-6]). A vector-valued model-based cumulative sum detection method was proposed [7]. The potential performance degradation that a stealthy attack can induce while remaining undetected was characterized. A generalized chi-squared detector for linear systems with non-Gaussian noise was proposed [8]. Three detection-related concepts were presented, including randomized switching of the control system to help detect attacks, a residual-based detection scheme using the difference between the measured state and predicted state as the residual, and a redundant state estimation-based detection scheme. Several models of cyberattacks were considered in [10]. Neural network-based detector schemes were developed and trained based on extensive simulation studies of the attack-free and attacked process [10]. While there has been a degree of acknowledgment that the control system design plays a role in the ability or inability to detect cyberattacks, the connection between control system design and cyberattack detectability has not been rigorously established.

In the present work, the role of controller design on cyberattack detectability is analyzed. Specifically, false-data injection attacks, attacks that manipulate the data communicated over the control system communication links, are considered. Rigorous definitions related to attack detectability are developed for a general class of detection schemes that encompass several commonly employed detection schemes. An analysis of attack detectability reveals that the control design impacts the ability to detect cyberattacks. These findings are verified using a chemical process where several detection schemes are considered under different control system designs. The results demonstrate that an attack may be detected or undetected depending on the control system design.

References

1. Hemsley, K., and Fisher, R., History of industrial control system cyberattacks, Idaho National Laboratory, Report number: INL/CON-18-44111, 2, 1-37 (2018).

2. Kayan, H., Nunes, M., Rana, O., Burnap, P., and Perera, C., Cybersecurity of industrial cyber-physical systems: A review, ACM Computing Surveys, (in press).

3. Ginter, A., The top 20 cyberattacks on industrial control systems, Technical Report, Version 1.1, Waterfall Security Solutions (2018).

4. Giraldo, J., Urbina, D., Cardenas, A., Valente, J., Faisal, M., Ruths, J., Tippenhauer, N. O., Sandberg, H., and Candell, R., A survey of physics-based attack detection in cyber-physical systems, ACM Computing Surveys, 51, 1-36 (2018).

5. Ding, D., Han, Q., Xiang, Y., Ge, X., and Zhang, X.. A survey on security control and attack detection for industrial cyber-physical systems, Neurocomputing, 275, 1674-1683 (2018).

6. Tan, S., J. M. Guerrero, P. Xie, R. Han, and J. C. Vasquez. Brief survey on attack detection methods for cyber-physical systems. IEEE Systems Journal, 14:5329-5339, 2020.

7. Murguia, C., Ruths, J., CUSUM and chi-squared attack detection of compromised sensors, Proceedings of the 2016 IEEE Conference on Control Applications, 474-480 (2016).

8. Hashemi, N., and Ruths, J., Generalized chi-squared detector for LTI systems with non-Gaussian noise, Proceedings of the 2019 American Control Conference, 404-410 (2019).

9.Oyama, H. and Durand, H., Integrated cyberattack detection and resilient control strategies using Lyapunov-based economic model predictive control, AIChE Journal, 66, e17084 (2020).

10. Chen, S., Wu, Z., and Christofides., P. D., Cyber-attack detection and resilient operation of nonlinear processes under economic model predictive control, Computers andamp; Chemical Engineering, 136, 106806 (2020).

419b Keynote Talk - Cybersecurity for Operational Technology and Smart Manufacturing Systems

CheeYee Tang

Many manufacturers and system operators use operational technology (OT) and industrial control system (ICS) to automate their operations. These systems are often operated in demanding environments and have unique performance, reliability, and safety requirements. Traditionally, these systems are air-gapped but an increasing number of operators are connecting their systems to the network to address business needs and boost productivity. Additionally, emerging technologies like IoT/IIoT and cloud computing are being integrated with these environments to support digital transformation and process optimization.

As more systems are interconnected, it is critical for these organizations to secure these systems. Typical IT security architectures like defense-in-depth and zero trust may require additional measures and considerations to work effectively in the OT and ICS environments. Unfortunately, many organizations are still having difficulty to fully implement cybersecurity measures to secure their infrastructures and OT systems.

NIST is developing standards and guidelines to assist organizations to address their cybersecurity needs. Publications like the NIST SP 800-82 provides guidance for OT systems and to demonstrate how to implement cybersecurity measures in the OT and ICS environments while addressing the performance and safety requirements of these systems.

419c Safe Control of Networked Chemical Process Plants Under Cyber-Attacks

Jaewon Kim, Akshay Mete, P. R. Kumar

Abstract

Cyber-Physical Systems (CPS) such as chemical process control plants, autonomous vehicles, unmanned aerial vehicles, and smart power systems are susceptible to a variety of cyber-attacks. These systems depend on the measurements reported from their sensors to close their control loops. This makes them susceptible to cyber-attacks where malicious agents can compromise the sensors or the networks carrying the sensor measurements. In recent years, there was have been several such cyber-attacks [1] [2] [3].

Our focus in this paper is on systems where the sensor measurements are compromised by a malicious agent at some point, either at the sensor or in the network transporting the sensor measurements which distorts the sensor measurements provided to the actuator in the control loop. Such cyber-attacks can be detected successfully by ``Dynamic Watermarking (DW)' methodology proposed in [4]. The basis and background of the DW method involves two specific statistical tests for security, and provides theoretical guarantees for linear stochastic systems. It has been shown to be capable of detecting cyber-attacks on chemical process control systems [5] [6], autonomous vehicles [7], unmanned aerial vehicles [8], and smart power systems [9]. However, the safe operation of the plant under attack remains an open problem. In this paper, we address the safe post-attack control of chemical process control plants.

Fig.1 show how the dynamic watermarking method can be used to trigger a switch which changes the feedback control loop. Prior to attack, the control loop employs the reported sensor measurements. Once the attack is detected by dynamic watermarking-based attack detector, the control loop is switched to so that it does not rely on the compromised sensor measurements. Rather it employs the estimated output of a model obtained by a system identification algorithm employed prior to attack.

Using the online system identification methods such as least squares estimation, we generate estimates of the sensor measurements under cyber-attack. We use these estimates to close the control loop rather than using malicious sensor measurements. Through both simulations and experimental validation we show the effectiveness of the proposed post-attack controller for safe control under cyber-attacks. The proposed post-attack controller along with the DW based Attack Detector thereby makes possible secure and safe operation of the chemical process control system under arbitrary cyber-attacks.

References

[1] Kerr, Paul K., John Rollins, and Catherine A. Theohary. The stuxnet computer worm: Harbinger of an emerging warfare capability. Washington, DC: Congressional Research Service, 2010.

[2] Farwell, James P., and Rafal Rohozinski. ``Stuxnet and the future of cyber war.' Survival 53.1 (2011): 23-40.

[3] Case, Defense Use. ``Analysis of the cyber attack on the Ukrainian power grid.' Electricity Information Sharing and Analysis Center (E-ISAC) 388 (2016): 1-23.

[4] Satchidanandan, Bharadwaj, and Panganamala R. Kumar. ``Dynamic watermarking: Active defense of networked cyber–physical systems.' Proceedings of the IEEE 105.2 (2016): 219-240.

[5] Kim, Jaewon, Woo-Hyun Ko, and P. R. Kumar. ``Cyber-Security with Dynamic Watermarking for Process Control Systems.' 2019 AIChE Annual Meeting. AIChE, 2019.

[6] Kim, Jaewon, and P. R. Kumar.``Security of Control Systems with Erroneous Observations.' IFAC-PapersOnLine 53.2 (2020): 2225-2230.

[7] Ko, Woo-Hyun, Bharadwaj Satchidanandan, and P. R. Kumar. ``Theory and implementation of dynamic watermarking for cybersecurity of advanced transportation systems.' 2016 IEEE Conference on Communications and Network Security (CNS). IEEE, 2016.

[8] Kim, Jaewon, Woo-Hyun Ko, and P. R. Kumar. ``Cyber-Security through Dynamic Watermarking for 2-rotor Aerial Vehicle Flight Control Systems.' 2021 International Conference on Unmanned Aircraft Systems (ICUAS). IEEE, 2021.

[9] Ramos-Ruiz, Jorge, et al. ``An active detection scheme for cyber attacks on grid-tied PV systems.' 2020 IEEE CyberPELS (CyberPELS). IEEE, 2020.

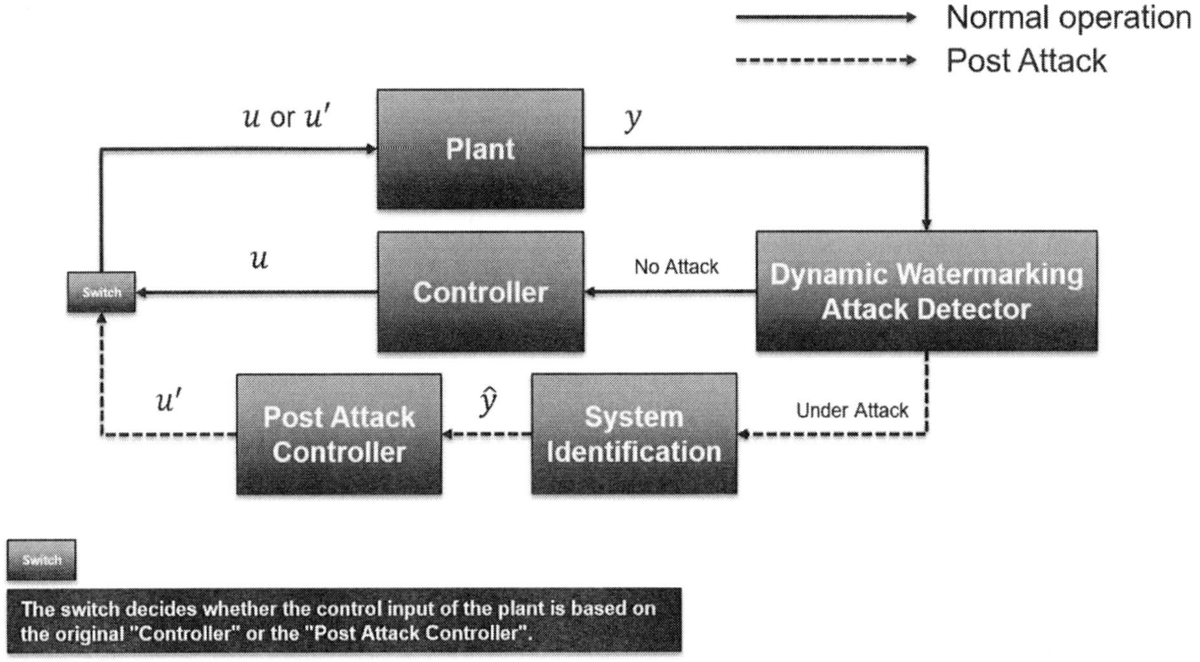

419d Lyapunov-Based Economic Model Predictive Control with Cyberattack Detection for Process Actuators

Keshav Kasturi Rangan, Henrique Oyama, Helen Durand

Recent years have seen industries working to more tightly integrate physical processes on process shop floors with communication and control systems [1]. This cyber-physical systems (CPS) integration is a consequence of advantages like increased production efficiency and improved transparency into the steps involved in production for better quantity control. However, this integration also opens the integrated processes and components to cyberattacks which can cause the degradation of process safety resulting in effects ranging from loss in profits to endangering human lives. Various aspects of a physical system can be implemented by an attack, but these different channels lead to actuators taking actions they would not have taken in the absence of an attack. However, there is a difference between having these actuators directly targeted via an attack (through, for example, improper control signals being sent to them), compared to having the sensors manipulated so that the actuators are receiving the correct control action for the state measurement that the controller received. Much of the prior work in our group on developing a control-theoretic framework for analyzing detection and control policies as they relate to cyberattacks on nonlinear systems has focused on attacks on the process sensors in the sense that the sensor measurements received by the controller are the part of the control loop that becomes in accurate.[3,4] These detection and control policies have focused on the use of an advanced optimization-based control framework known as Lyapunov-based economic model predictive control (LEMPC) [2] in the detection of attacks when the actuators are assumed to function correctly. However, when they function incorrectly, it is necessary to consider how these strategies might be redesigned.

This work discusses our advances in integrated cyberattack detection and control policies when actuators are impacted.[5,6] We first discuss how three integrated cyberattack detection and control policies from [4] can be modified to for the case that actuators are attacked instead of sensors. The first of these methods involves constantly probing for cyberattacks by developing new steady-states throughout state-space over time where the closed-loop state measurements must show that the state is approaching the new steady-state during operation at all times. The second method involves using predictions of the state under the expected control action and comparing this with the state measurements observed. The third involves using redundant state estimators to catch when the closed-loop state estimates are inconsistent. We demonstrate closed-loop stability under the first two of these strategies in the actuator attack case, and discuss challenges with implementing the third. We also discuss how these strategies relate to profits and to what aspects of a system would need to be fully secured to gain detection guarantees. The implementation strategy of the constant probing strategy is demonstrated through a continuous stirred tank reactor example. This example serves to showcase that part of the implementation strategy must involve how to ensure that the alternative steady-states developed over time are steady-states within the input bounds.

References:

[1] Lezzi, M., Lazoi, M., Corallo, A. Cybersecurity for industry 4.0 in the current literature: A reference framework. Computers in Industry (103), 97-110 (2018)

[2] Heidarinejad, M., Liu, J. and Christofides, P.D. Economic model predictive control of nonlinear process systems using Lyapunov techniques. AIChE Journal, 58(3), 855-870 (2012).

[3] Rangan, K. K., Oyama, H., andamp; Durand, H. Integrated Cyberattack Detection and Handling for Nonlinear Systems with Evolving Process Dynamics under Lyapunov-based Economic Model Predictive Control. Chemical Engineering Research and Design (2021).

[4] Oyama, H. and H. Durand. Integrated Cyberattack Detection and Resilient Control Strategies Using Lyapunov-Based Economic Model Predictive Control, AIChE Journal (66), ee17084 (2020).

[5] Rangan, K. K., Oyama, H., andamp; Durand, H. Actuator Cyberattack Handling Using Lyapunov-based Economic Model Predictive Control. 13th IFAC Symposium on Dynamics and Control of Process Systems, including Biosystems (DYCOPS 2022).

[6] Oyama, H., Rangan, K. K., andamp; Durand, H. Lyapunov-Based Economic Model Predictive Control for Detecting and Handling Actuator and Simultaneous Sensor/Actuator Cyberattacks on Process Control Systems. Frontiers in Chemical Engineering, 11.

419e Switching-Enabled Active Detection of False-Data Injection Cyberattacks on Process Control Systems

Shilpa Narasimhan, Nael El-Farra, Matthew Ellis

In the past decade, cyberattacks targeting industrial control systems such as process control systems (PCSs) have increased in frequency and sophistication [1]. To bolster the cybersecurity protections provided by information technology infrastructure, a variety of controller-based approaches for cybersecurity have been proposed [2]-[5]. Cyberattack-resilient control approaches [2], attack detection schemes (e.g., [3]-[5]), and attack identification and mitigation strategies (e.g., [6], [7]) are a few examples of areas being explored.

Several classes of cyberattacks targeting PCSs have been considered [8]. False data injection (FDI) attacks are a class of attacks that are designed to alter the data communicated over the communication links of a PCS. Two types of FDI attacks that have been considered are additive attacks that alter data by adding a factor to the data being communicated [3] and multiplicative attacks that alter data by multiplying a factor to the data [4], [5]. Compared to additive FDI attacks, multiplicative FDI attacks may require minimal process knowledge to alter process operations, while evading detection [10].

Cyberattack detection schemes may be broadly classified as either passive or active detection schemes. Passive detection schemes monitor a PCS for attacks based on regular operational data [3]. Residual-based detection schemes are commonly used anomaly detection schemes for process monitoring [9] that have been employed for detecting cyberattacks [3], [10]. Residual-based detection schemes detect an attack if an anomaly is observed in residual values (defined as the difference between the measured output and its expected value). Because they monitor processes based on regular operational data, residual-based detection schemes are considered passive schemes. Active detection schemes, on the other hand, actively probe the process for cyberattacks by applying external interventions or perturbations. For example, an active detection method may utilize the addition of a secret watermarking signal to the sensor data or to the actuator data to detect an attack (e.g., [4], [5]).

In this work, a control parameter switching-based active detection methodology that enhances the detection capabilities of a residual-based detection scheme is proposed. The proposed switching-based active detection methodology is motivated by the intrinsic connection between the control parameter selection, the closed-loop stability, and the detectability of an attack with respect to a residual-based detection scheme. To this end, this relationship is rigorously analyzed to identify the set of control parameters (called "attack-sensitive" parameters) under which an attack renders the closed-loop system unstable and can therefore be detected. Extended operation under attack-sensitive parameters, however, may result in undesired performance degradation of the attack-free closed-loop system. To manage the trade-off between attack detection enhancement and closed-loop performance degradation, the proposed active detection methodology utilizes occasional control parameter switching from nominal parameters (chosen based on standard control design criteria) to the attack-sensitive parameters. Control parameter switching may excite the process dynamics, resulting in false alarms. Therefore, a switching condition is developed to minimize false alarms. The proposed active detection methodology is applied to a chemical process example and shown to enhance attack detection capabilities, beyond those achieved by passive detection schemes, while also minimizing false alarms.

References:

[1] T. Miller, A. Staves, S. Maesschalck, M. Sturdee, and B. Green, "Looking back to look forward: Lessons learnt from cyber-attacks on industrial control systems," International Journal of Critical Infrastructure Protection, 35:100464, 2021.

[2] H. Oyama and H. Durand, "Integrated cyberattack detection and resilient control strategies using Lyapunov-based economic model predictive control," AIChE Journal, 66(12):e17084, 2020.

[3] C. Murguia and J. Ruths, "Characterization of a CUSUM model-based sensor attack detector," In Proceedings of the IEEE 55th Conference on Decision and Control, pp. 1303–1309, Las Vegas, NV, USA, 12-14 December 2016.

[4] T. Huang, B. Satchidanandan, P. R. Kumar, and L. Xie, "An online detection framework for cyber attacks on automatic generation control," IEEE Transactions on Power Systems, 33:6816–6827, 2018.

[5] G. Na and Y. Eun, "A multiplicative coordinated stealthy attack and its detection for cyber physical systems", In Proceedings of the IEEE Conference on Control Technology and Applications, pp. 1698–1703, Copenhagen, Denmark, 21-24 August 2018.

[6] M. Kordestani and M. Saif, "Observer-based attack detection and mitigation for cyberphysical systems: A review," IEEE Systems, Man, and Cybernetics Magazine, 7(2):35–60, 2021.

[7] Zedan, A. and N. H. El-Farra, "A machine-learning approach for identification and mitigation of cyberattacks in networked process control systems," Chemical Engineering Research and Design, 176: 102-115, 2021.

[8] C. Xenofontos, I. Zografopoulos, C. Konstantinou, A. Jolfaei, Muhammad K. Khan, and K. K. R. Choo, "Consumer, commercial, and industrial IoT (In)security: Attack taxonomy and case studies," IEEE Internet of Things Journal, 9:199–221, 2022.

[9] M. Blanke, M. Kinnaert, J. Lunze, M. Staroswiecki, "Diagnosis and Fault-Tolerant Control," Springer-Verlag Berlin Heidelberg, 2003.

[10] S. Narasimhan, N. H. El-Farra, and M. J. Ellis, "Detectability-based controller design screening for processes under multiplicative cyberattacks," AIChE Journal, 68:e17430, 2022.

419f Modeling Impacts of Cyberattacks on Control of Powder Bed Fusion

Kip Nieman, Helen Durand

Next-generation manufacturing will involve new and innovative methods of manufacturing, requiring increased flexibility. Additive manufacturing methods (i.e., 3D printing) are enticing as they can handle many different configurations for producing unique and complex structures, including for prototyping and manufacturing, in a wide variety of fields [1, 2]. Powder bed fusion (PBF) is one kind of additive manufacturing method that can be used to produce metal objects through selectively melting powder using a laser. Between each layer, a roller or sweeping mechanism spreads a fresh layer of powder over the previously melted layer. The laser then fuses the next layer to the previous, and successive layers are added to create the final completed part. The advantage of PBF is that high-quality metal parts can be produced, but post-processing is typically resource and labor-intensive. This includes ensuring part quality, finishing, and heat treatment [3]. Current methods of improving part quality (and thus minimizing post-processing) usually involve tuning production parameters based on previous experience or testing [3, 4], or though control of the laser or laser melt pool [5]. Advanced control methods have been used for tracking a desired melt pool behavior [5]. However, as control concepts for additive manufacturing become more sophisticated, it becomes necessary to consider how attacks on control systems might impact additively manufactured parts. It has been recognized that a potential challenge for additive manufacturing is cyberattacks on the control systems [6], which could, for example, introduce defects in a part.

In [7, 8], we discussed simulation of rogue input policies in the context of a steam methane reformer (SMR) where both transport and equipment stress considerations were important; it would be expected that similar studies could aid in showcasing the impacts of attacks on additively manufactured parts. While both processes can be simulated in ANSYS with finite element methods, they consider different phenomena that require unique considerations. In particular, PBF must consider the evolving part shape which could be manipulated in many ways, including void creation, incomplete melting, warping or cracking, or altered dimensions.

This talk will build from our prior studies using ANSYS for simulating rogue control policies in the context of the more traditional chemical process of steam methane reforming to focus on additive manufacturing, and the criticality of understanding the effects of rogue control actions on additively manufactured parts. We will discuss our progress on considering cybersecurity within an additive manufacturing process simulation [9] where both transport phenomena (in this case, heat transfer) and stress in a part are important. We will discuss modeling of the powder bed fusion process with a single powder layer on a baseplate, where the process is modeled with finite element analysis (FEA) in ANSYS based on modeling strategies synthesized from [10-13]. The process was simulated using a three-dimensional geometry and through a coupled ANSYS thermal and structural analysis. ANSYS parametric design language (APDL) code was written to apply volumetric heat generation (based on [12]) to represent the temperature change induced by the laser, to manage the phases involved (powder, solid, and liquid), and to manage the coupling of the thermal and structural simulations. This code also enables, for example, the velocity of the laser to be adjusted to see how changes in this type of control input can impact the part (both in terms of the part geometry and stress profiles) and to suggest strategies for detecting when attacks that might impact these inputs occur. We discuss how different mesh sizes that are not necessarily at the level of mesh independence impact what can be learned from cyberattack studies, in particular in the case that major changes to the dynamics of the system occur (e.g., if the laser moves at a different speed), to better understand the computational resources required for a full analysis of cybersecurity concerns for additive manufacturing simulation.

REFERENCES:

[1] J. Plocher, and P. Ajit. 'Review on design and structural optimisation in additive manufacturing: Towards next-generation lightweight structures.' Materials andamp; Design 183 (2019): 108164.

[2] J. Chang, J. He, M. Mao, W. Zhou, Q. Lei, X. Li, D. Li, C.K. Chua, and X. Zhao. 'Advanced material strategies for next-generation additive manufacturing.' Materials 11.1 (2018): 166.

[3] H.M. Khan, Y. Karabulut, O. Kitay, Y. Kaynak , and I. S. Jawahir. 'Influence of the post-processing operations on surface integrity of metal components produced by laser powder bed fusion additive manufacturing: a review.' Machining Science and Technology 25.1 (2020): 118-176.

[4] N. Ahmed, I. Barsoum, and G. Haidemenopoulos. 'Process parameter selection and optimization of laser powder bed fusion for 316L stainless steel: A review.' Journal of Manufacturing Processes 75 (2022): 415-434.

[5] M. Mani, B.M. Lane, M.A. Donmez, S.C. Feng, and S.P. Moylan. 'A review on measurement science needs for real-time control of additive manufacturing metal powder bed fusion processes.' International Journal of Production Research 55.5 (2017): 1400-1418.

[6] S. Y. Yu, A. V., Malawade, S. R. Chhetri, and M. A. Al Faruque. 'Sabotage attack detection for additive manufacturing systems.' *IEEE Access* 8 (2020): 27218-27231.

[7] K. Nieman, H. Oyama, M. Wegener, and H. Durand. 'Predict the Impact of Cyberattacks on Control Systems.' *Chemical Engineering Progress*. Sept. 2020.

[8] K. Nieman, D. Messina, M. Wegener, and H. Durand. 'Cybersecurity and Dynamic Operation in Practice: Equipment Impacts and Safety Guarantees.' *Journal of Loss Prevention in the Process Industries*. Submitted Feb. 15, 2021.

[9] K. Nieman, Leonard, A. F., K. Tyrrell, D. Messina, R. Lopez, H. Durand. "Challenges and Opportunities for Next-Generation Manufacturing in Space." International Federation of Automated Control (IFAC) Symposium on Dynamics and Control of Process Systems (DYCOPS), In Press 2022.

[10] A. Hussein, L. Hao, C. Yan, and R. Everson. 'Finite element simulation of the temperature and stress fields in single layers built without-support in selective laser melting.' *Materials andamp; Design* (1980-2015) 52 (2013): 638-647.

[11] K. Zeng, D. Pal, H. Gong, N. Patil, and B. Stucker. 'Comparison of 3DSIM thermal modelling of selective laser melting using new dynamic meshing method to ANSYS.' *Materials Science and Technology* 31.8 (2015): 945-956.

[12] J. Goldak, A. Chakravarti, and M. Bibby. 'A new finite element model for welding heat sources.' *Metallurgical transactions B* 15.2 (1984): 299-305.

[13] D. Deng and K. Shoichi. 'Numerical simulation of residual stresses induced by laser beam welding in a SUS316 stainless steel pipe with considering initial residual stress influences.' *Nuclear Engineering and Design* 240.4 (2010): 688-696.

419g Limitations of Control-Theoretic Control System Cyberattack Detection for Distributed Control Leading to Cost/Benefit Analysis for Control System Cybersecurity

Dominic Messina, Katrina Hinzman, Helen Durand

In recent years, there have been growing concerns regarding the cybersecurity of chemical plants in the wake of numerous successful attacks on control and information technology (IT) systems, resulting in loss of profits and destruction of property [2], as well as the potential for casualties if potentially dangerous processes are targeted in the future. The obstacles for cybersecurity of industrial control systems (ICS) are twofold: the sheer amount of potential attack surfaces (e.g. hardware components, software, and communication networks [3]) requires careful integration and strategies for each type of attack, while process operation is considered very high priority, whereas in many other cases the security of the system takes precedence over operational availability [1]. Therefore, it is of interest to develop secure ICS architectures capable of handling cyberattacks on all system components with the interest of sufficiently guaranteeing safe continuous process operation at the lowest possible cost. In this talk, we will discuss cyberattack handling for attacks on ICS hardware components (e.g. sensors) as well as IT considerations for network communication security.

Cyberattacks targeting hardware components and software have potential to cause adverse events resulting in the loss of control of a process. Detection and removal of a cyberattack is critical for keeping processes online. In the case of centralized control, where a single controller operates the process, detection methods have been developed to flag cyberattacks on a system, however to direct attack removal methods it is important to diagnose which specific components are being attacked (e.g. a compromised thermocouple which is sending false temperature readings). Distributed control frameworks, which utilize communicating networks of controllers operating on process subsystems and are potentially attractive for reducing computation time of control actions for large scale processes, have an increased number of components and communication links that could facilitate their attack. However, distributed control also has potential to operate processes safely in the event that a controller or subsystem must be taken offline (due to, for example, an attack diagnosed on a certain part of the system), making cyberattack diagnosis critical for continuous operation. Thus, distributed control seems as though it may be attractive for applications in cyberattack resilient ICS frameworks.

Despite the seeming promise of distributed control architectures for handling cyberattacks on control systems, we discuss how the extension of the integrated cyberattack detection and control strategies in [4] can proceed in a straightforward manner. We discuss a variety of concepts for attempting to cause the larger number of controllers to cross-check one another to attempt to obtain a degree of agreement among controllers that the system appears to be functioning appropriately. However, attempts to diagnose and validate the operation of the system through cross-checking can add cost due to redundancy and complexity, and showcase the value in seeking solutions to the cyberattack diagnosis problem that go beyond breaking the system up into many cross-validating components. We also discuss closed-loop stability considerations which govern when removal of attacked distributed control components may be feasible, and what those conditions imply about how attackers could gain the upper hand over the distributed detection system by taking control of it when it is in a condition from which removal of a component does not permit stabilization by the remaining controllers when the attacked actuators have a certain policy. We conclude by discussing how our work fits within the context of information technology-based approaches, and how one might begin to think about comparing the use of control-theoretic cybersecurity measures with information technology approaches, with cost/benefit analyses for pairings of different technologies.

[1] Candell, Richard, et al. 'A cybersecurity testbed for industrial control systems.' *Proceedings of the 2014 Process Control and Safety Symposium*. (2014).

[2] Iaiani, Matteo, et al. 'Analysis of past cybersecurity-related incidents in the process industry and the like.' *Chemical Engineering Transactions* 82 (2020): 163-168.

[3] McLaughlin, Stephen, et al. 'The cybersecurity landscape in industrial control systems.' *Proceedings of the IEEE* 104.5 (2016): 1039-1057.

[4] Oyama, Henrique, and Durand, Helen. 'Integrated cyberattack detection and resilient control strategies using Lyapunov-based economic model predictive control.' *AIChE Journal* 66.12 (2020): e17084.

286a Fusion of EEG and Eye-Tracking Based Metrics for Characterizing the Cognitive State of Control Room Operators

Mohd Umair Iqbal, Babji Srinivasan, Rajagopalan Srinivasan

Process industries are highly hazardous, and the continued occurrence of accidents points toward a need to develop more robust and resilient safety barriers. Several layers of protection prevent hazards from transforming into accidents, and humans form one of the most critical layers. The importance of humans as a layer of protection can be envisioned by the fact that most of the accidents in process industries trace their roots to human failures. Human errors account for over 60-80% of accidents in process industries (Brauer, 2016). An analysis of severe industrial accidents from 2008 to 2018 reveals that more than 76.1% can be attributed to human failures (Jung et al., 2020). With an increased digitalization and concomitant rise in automation in process industries, the role of humans has evolved. The human roles primarily involve monitoring during routine operations and diagnosis and prognosis during abnormalities. These roles have become cognitively challenging due to a need to acquire, segregate, and integrate information from multiple sources to maintain optimal operating conditions. A requirement to process an overabundance of information increases distraction and cognitive workload— precursors to suboptimal performance— if the operators' capabilities do not match process demands (Porthin et al., 2021). This paper addresses the critical need to understand and best evaluate cognitive aspects of control room operators' performance using a fusion of eye-tracking and electroencephalography (EEG)-based metrics.

Traditional methods of human performance assessment ignore cognitive aspects of operators' performance. On the contrary, the performance is assessed primarily based on metrics such as the number of successes/failures, deviations of operator actions from standard operating procedures, and response time (Lee et al., 2000). Such metrics fail to account for the underlying reasons (e.g., cognitive workload, shortcomings in mental models) and contributory factors (e.g., the effect of HMI design) to sub-optimal performance. Advancements in cognitive engineering and concurrent development of sensor technologies such as eye-tracking and electroencephalography (EEG) have made it possible to obtain several physiological measures that provide critical insights into human cognition and performance. Eye-tracking primarily provides information about how an operator acquires information from HMIs (reflection of operator's mental models) (Shahab et al., 2022) and how HMI designs influence performance (Diego-Mas et al., 2019). Likewise, electroencephalography (EEG) provides critical information about operators' mental states, such as levels of cognitive workload (Liang et al., 2018). Nonetheless, eye-tracking alone provides limited insight into the mental processes; brain-sensing can complement these processes. For instance, a classification accuracy of 93% for visual hazard detection was attained when features from eye-tracking and EEG were fused compared to accuracies of 73% (for eye-tracking) and 83% (for EEG) obtained independently (Noghabaei et al., 2021). Our previous research using eye-tracking (Sharma et al., 2016, Bhavsar et al., 2017) and EEG (Iqbal et al., 2020, 2021) reveals exciting insights into the cognitive behavior of control room operators. For instance, experts have more directed attention as captured by lower gaze entropy than novices (Bhavsar et al., 2017). EEG is able to identify instances when there is a mismatch between the operator's mental models of process and actual process behavior (Iqbal et al., 2020).

In the current work, we study the added benefit of the fusion of eye-tracking and EEG for operator performance assessment. The proposed approach for fusion involves identifying key events during a typical period during which an operator interacts with the process. These key events could include: the periods during which an operator is simply monitoring the process when it runs normally; periods which involve disturbances; and periods during which an operator is actively taking control actions to bring the process back within normal operating conditions. The purpose of the approach is to identify which metrics—eye-tracking and/or EEG-based—can best capture operator's cognition during these events, and provide insights into their cognitive performance. For instance, whether eye-tracking based metrics are more sensitive to performance (say, level of attention) during monitoring or EEG-based. Likewise, which metrics can quickly capture increase in cognitive workload in the wake of inability of an operator to control process abnormalities, and which respond after a lag. To obtain answers to these questions, we use a simulated ethanol production process as an experimental testbed for conducting human subject studies. Ten participants were involved in the study, and each of them carried out several repetitive trials. During a typical trial, a participant has to carry out six tasks (each corresponding to a disturbance scenario). Overall, these participants performed 81 trials resulting in 438 task-participant pairs. During these tasks, participants' brain activity and eye gaze are recorded via a single electrode EEG sensor and Tobii 300 eye tracker, respectively. In addition, process data, alarm information, and operator actions are also recorded. These are used to create labels such as the monitoring phase, the period immediately after an alarm, before an operator's action, after an operator's first slider action, and the period during which the operator executes a control action. Next, eye-tracking and EEG-based features are extracted for each of these labels. These features include average pupil diameter, power spectral densities in the range of 0 to 2 Hz for pupil diameter, and power spectral densities in the range of 0-30 Hz for EEG. Decision tree-based analysis of features reveals that combining features from eye and EEG data provides more insights into operator cognitive performance than either of these. This is hypothesized due to differences in dynamics (e.g., timescale) of EEG and eye data. In this paper, we will present the proposed fusion methodology and present results from our human subject studies that demonstrate the benefits of the fusion of eye tracking and EEG based metrics.

Acknowledgement: This work is partially funded by American Express Lab for Data Analytics,

Risk and Technology, Indian Institute of Technology Madras

References

Bhavsar, P., Srinivasan, B., andamp; Srinivasan, R. (2017). Quantifying situation awareness of control room operators using eye-gaze behavior. Computers andamp; chemical engineering, 106, 191-201.

Brauer, R. L. (2016). Safety and health for engineers. John Wiley andamp; Sons.

Diego-Mas, J. A., Garzon-Leal, D., Poveda-Bautista, R., andamp; Alcaide-Marzal, J. (2019). User-interfaces layout optimization using eye-tracking, mouse movements and genetic algorithms. Applied ergonomics, 78, 197-209.

Iqbal, M. U., Shahab, M. A., Choudhary, M., Srinivasan, B., andamp; Srinivasan, R. (2021). Electroencephalography (EEG) based cognitive measures for evaluating the effectiveness of operator training. Process Safety and Environmental Protection, 150, 51-67.

Iqbal, M. U., Srinivasan, B., andamp; Srinivasan, R. (2020). Dynamic assessment of control room operator's cognitive workload using Electroencephalography (EEG). Computers andamp; Chemical Engineering, 141, 106726.

Jung, S., Woo, J., andamp; Kang, C. (2020). Analysis of severe industrial accidents caused by hazardous chemicals in South Korea from January 2008 to June 2018. Saf. Sci. 124, 104580

Lee, S., Jeong, I., andamp; Il, M. (2000). Development of evaluation algorithms for operator training system. Computers andamp; Chemical Engineering, 24(2-7), 1517-1522.

Liang, Y., Liang, W., Qu, J., andamp; Yang, J. (2018, October). Experimental study on EEG with different cognitive load. In 2018 IEEE International Conference on Systems, Man, and Cybernetics (SMC) (pp. 4351-4356). IEEE.

Noghabaei, M., Han, K., andamp; Albert, A. (2021). Feasibility Study to Identify Brain Activity and Eye-Tracking Features for Assessing Hazard Recognition Using Consumer-Grade Wearables in an Immersive Virtual Environment. Journal of Construction Engineering and Management, 147(9), 04021104.

Porthin, M., Liinasuo, M., andamp; Kling, T. (2020). Effects of digitalization of nuclear power plant control rooms on human reliability analysis–A review. Reliability Engineering andamp; System Safety, 194, 106415.

Shahab, M. A., Iqbal, M. U., Srinivasan, B., andamp; Srinivasan, R. (2022). HMM-based models of control room operator's cognition during process abnormalities. 1. formalism and model identification. Journal of Loss Prevention in Process Industries, 104748.

Sharma, C., Bhavsar, P., Srinivasan, B., andamp; Srinivasan, R. (2016). Eye gaze movement studies of control room operators: A novel approach to improve process safety. Computers andamp; Chemical Engineering, 85, 43-57.

286b Development and Application of Customized Symbols in PI Vision for KPI Monitoring Based on the Greenscope Methodology

Esmael Gadelha, Igor Guerra, Natalya A. B. Almeida, Beatriz Dantas, Fernando V. Lima, Heleno Bispo

The fourth industrial revolution is reshaping the way of working for professionals and companies. In this revolution, the use of technologies based on data aims to supply information to improve production processes. This is possible due to the increase of MES (Manufacturing Execution System) and ERP (Enterprise Resource Planning) systems in industrial plants. Thus, the integration of techniques and engineering procedures into such systems may help the process production decisions in the context of sustainability. Sustainable development can be reached through a new vision attributed to industrial processes in many areas, such as efficiency, energy, economy, and environment. For example, the determination of process conditions to control gas emissions, reuse, and recycling of waste [1]. In this way, GREENSCOPE emerges as a methodology to help the quantification of the process sustainability indices based on KPIs (Key Performance Indicators) [2].

This work is related to the monitoring of process indicators based on GREENSCOPE using the PI System Architecture. The following indicators have been selected for evaluating and monitoring a dynamic simulated chemical process: reaction yield, total energy cost, total energy consumption, resource energy efficiency, and environmental quotient [2, 3]. For this purpose, a PI System Client Tool, PI Vision, is used for visualization of the KPIs, via the PI AF (Asset Framework), as shown in Figure 1.

Through the PI Data Archive provided by the PI server, the chemical process data is processed on PI AF, and the KPIs are calculated. The visualization has been possible by the PI Vision, where an extensibility model allowed the development of custom symbols for a better understanding of the relation between the KPI values. Such symbols are written in HTML, CSS, and JavaScript languages, enabling the inclusion of exclusive/sector-specific forms of data visualization. For this purpose, a radar diagram symbol has been developed, allowing the monitoring of the performance of the implemented indicators in PI AF, providing robust critical analysis and assessment of the process in focus.

REFERENCES

1. Krajnc, D.; Glaviˇc, P. Indicators of sustainable production. Clean Technol. Environ. Policy 2003, 5, 279–288.
2. Ruiz-Mercado, G.; Smith, R. L.; Gonzalez, Appendix C: GREENSCOPE: Technical User's Guide. U. S. Environmental Protection Agency, 2014.
3. Ruiz-Mercado, G. J. et al. Sustainability Indicators for Chemical Processes: I. Taxonomy. Industrial andamp; Engineering Chemistry Research, 2011.

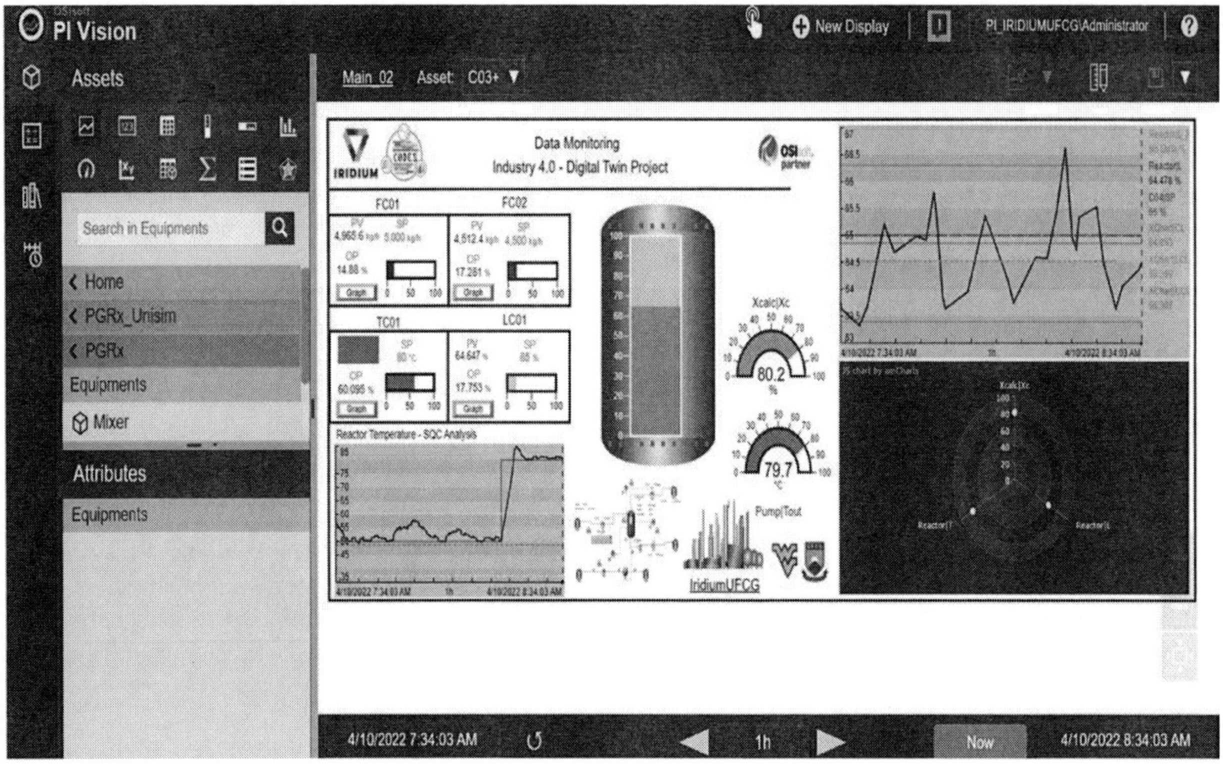
PI Vision
New Display
PI_IRIDIUMUFCG\Administrator
Assets
Search in Equipments
Home
PGRx_Unisim
PGRx
Equipments
Mixer
Attributes
Equipments
Main_02
Asset: C03+
Data Monitoring
Industry 4.0 - Digital Twin Project
IRIDIUM
FC01
FC02
TC01
LC01
Graph
Reactor Temperature - SQC Analysis
Xcalc|Xc
80.2
79.7
Pump|Tout
IridiumUFCG
4/10/2022 7:34:03 AM
1h
Now
4/10/2022 8:34:03 AM

286c Design Space Exploration and Optimization for Additive Manufacturing through Rational Feature Engineering and Machine Learning

Alexander Summers, Q. Peter He

Additive manufacturing is an emerging technology in which parts are fabricated layer-by-layer from computer-aided design models. Oxide dispersion strengthened (ODS) steels, the material subject of this work, are known for their high strength and resistance to radiation damage, making them ideal for structural materials in aerospace applications and in advanced fission and fusion reactors [1-2]. Pulse wave laser emission, as compared to continuous wave emission, can be used to improve surface quality and refine microstructure of the manufactured ODS steel parts, but it also leads to a much-increased design parameter space [3-5]. In this work, we perform rational feature engineering to find features that allow us to accurately and robustly classify "good" and "bad" quality parts. We then extend this exploration into a regression problem, using optical density and nano-indentation values. Finally, we determine the parameter ranges that consistently produce "good" quality, high-strength parts and describe the physical phenomena that explain these findings.

[1] S. Ukai, S. Ohtsuka, T. Kaito, Y. de Carlan, J. Ribis, J. Malaplate, Oxide dispersion-strengthened/ferrite-martensite steels as core materials for Generation IV nuclear reactors, Struct. Mater. Gener. IV Nucl. React. (2017) 357–414.

[2] H. Tanigawa, E. Gaganidze, T. Hirose, M. Ando, S.J. Zinkle, R. Lindau, E. Diegele, Development of benchmark reduced activation ferritic/martensitic steels for fusion energy applications, Nucl. Fusion. 57 (2017) 092004.

[3] S. Li, H. Xiao, K. Liu, W. Xiao, Y. Li, X. Han, J. Mazumder, L. Song, Melt-pool motion, temperature variation and dendritic morphology of Inconel 718 during pulsed- and continuous-wave laser additive manufacturing: A comparative study, Mater. Des. 119 (2017) 351–360.

[4] Y. Tian, J.A. Muñiz-Lerma, M. Brochu, Nickel-based superalloy microstructure obtained by pulsed laser powder bed fusion, Mater. Charact. 131 (2017) 306–315.

[5] C. Guo, Y. Zhou, X. Li, X. Hu, Z. Xu, E. Dong, Q. Zhu, R. Mark Ward, A comparing study of defect generation in IN738LC superalloy fabricated by laser powder bed fusion: Continuous-wave mode versus pulsed-wave mode, J. Mater. Sci. Technol. 90 (2021) 45–57.

286d Images, Motion Capture, Three-Dimensional Modeling, and Haptics for Future Manufacturing

Govanni Gjonaj, Renee O'Neill, Minhazur Rahman, Jacob Headley, Paloma Beacham, Samantha Cherney, Michael Williamson, Keshav Kasturi Rangan, Helen Durand

Many advances related to the field of computer graphics are ushering in a large number of ideas for the future of manufacturing. For example, virtual reality has been discussed in the context of training [1]. However, as a large number of advances for automation and efficiency are contemplated for the process industries, it becomes interesting to consider how one might test and evaluate whether these technologies will be beneficial for a given process, or otherwise to test how they might be designed to integrate effectively with other considerations at a plant such as humans, robotic systems, or artificially intelligent decision-making systems.

This talk presents case studies covering a variety of topics of relevance to next-generation manufacturing that fall broadly within contexts of relevance to computer graphics. Our goal is to specifically investigate how simulations of future manufacturing systems might be carried out. To reduce the complexity of these initial investigations, we perform the studies within the context of the open-source computer graphics software toolset Blender that has capabilities for image generation, Python scripting, and animation to facilitate the creation of scenes involving human-plant interaction, processes where images might play an integral role in control, or where the three-dimensional plant context is important to taking actions [2]. The first study to be presented will focus on how Blender might be used to create an image-based control simulation for a zinc flotation process. This process has previously been analyzed in an image-based control context (e.g., [3]). We first discuss how Blender can be used for simulating the visual movement of bubbles in the flotation process by modeling equations of velocity and position in the Python programming interface and using these to manipulate spheres in the 3D viewport with keyframes used to set the animation. We demonstrate what a render of the froth then looks like to showcase how Blender can be used for generation of the image to be used in the control design. We describe how the Python programming interface can be used to access images with tools such as OpenCV [4] that then allow operations to be performed on the images that could aid in image-based control. We then discuss how the model must be further modified by synthesizing information and inspiration from works which model both the physics (without graphics) of a flotation process [5] and those which model bubbles without flotation physics [6].

We then discuss several other advances that can be used to assess the use of Blender. For example, we showcase how chemical engineers might take plant data and reference images to seek to build the basic layout of a plant, and then also discuss how Blender can be used as a test framework for image-based safety monitoring algorithms that could generate many potential actions of a human when no plant data is available (as would be the typical case in an academic setting), and then use this data as if it was captured data from a plant to assess algorithms such as motion capture integrated with safety assessment. We also present some perspective on how evaluating haptic considerations might be evaluated with the aid of Blender.

[1] Fracaro, S. G., Chan, P., Gallagher, T., Tehreem, Y., Toyoda, R., Bernaerts, K., Glassey, J., Pfeiffer, T., Slof, B., Wachsmuth, S., andamp; Wilk, M. (2021). Towards design guidelines for virtual reality training for the chemical industry. *Education for Chemical Engineers*, *36*, 12-23.

[2] Oyama, H., D. Messina, R. O'Neill, S. Cherney, M. Rahman, K. K. Rangan, G. Gjonaj, and H.

Durand, "Test Methods for Image-Based Information in Next-Generation Manufacturing," *Proceedings of the IFAC Symposium on Dynamics and Control of Process Systems, including Biosystems (DYCOPS)*, Busan, Republic of Korea, in press.

[3] Kaartinen, J., Hätönen, J., Hyötyniemi, H., andamp; Miettunen, J. (2006). Machine-vision-based control of zinc flotation—a case study. *Control Engineering Practice, 14*(12), 1455-1466.

[4] Bradski, G. (2000). The OpenCV Library. *Dr. Dobb's Journal of Software Tools.*

[5] Do, H. (2010). *Development of a turbulent flotation model from first principles* (Doctoral dissertation, Virginia Tech).

[6] Cleary, P. W., Pyo, S. H., Prakash, M., andamp; Koo, B. K. (2007). Bubbling and frothing liquids. In *ACM SIGGRAPH 2007 papers* (Article 97).

286e Image Prediction for Model Predictive Control

Dominic Messina, Helen Durand

Autonomous robots equipped with vision and sensing have recently come to be implemented in chemical plants for process monitoring and inspection (such as the ANYmal robot [2]), with the expectation that human-robot interaction will be commonplace in next-generation manufacturing plants [7]. Machine learning based approaches are used extensively to help robots achieve tasks, such as movement [5], path planning [6], object detection [1], and more. While generally favorable in terms of computation times, machine learning models require a "black box" process model lacking insights necessary to interpret why actions are being chosen, which may be especially detrimental in contexts where flawless operation is crucial for safety or the completion of mission objectives (e.g. inspection of a plant after a disaster where hazardous materials may be present) [4]. Object detection in particular is a critical aspect (YOLOv4 [1] is an example); machine learning models are trained to detect and classify objects with a confidence interval in a robot's field of vision, which may be particularly useful for inspecting places where it is difficult for humans to go. However, due to the black box model and confidence interval, it may be beneficial to supplement object detection with an image-based control methodology utilizing predictive control that can steer the robot in ways which could increase its confidence in its assessment of its environment to promote safe and accurate performance.

In this talk, we discuss a concept for image prediction for integration with predictive control to investigate how and whether predictive control algorithms could be useful in a chemical process context if they seek to select movements for an agent that would aid it in distinguishing between different possible scenarios with respect to its environment. The premise of these studies is that an agent (e.g., robot) has developed several alternative possibilities of what it considers its environment might look like, but cannot with the available data fully come to a conclusion on what its environment looks like. It needs to gather new data to determine this; however, we would like it to achieve a control objective while gathering this new data (e.g., to gather this data in a manner that allows the agent to come to a conclusion about which of the possibilities is accurate as quickly as possible). We consider that there are directions that the robot could move that could enable it to figure out this information quickly, and others where the different options for what the environment looks like would continue to not be as readily distinguishable. We would like to design a strategy by which the robot takes a path by which it may most readily determine which of the visual models of its environment is the accurate one.

The first step in moving toward this goal is to select a simulation framework for images and analyze cases where, if a three-dimensional environment was to form the process model in a predictive controller, different movements of the agent around these environments would demonstrate different visual cues that could be used to select how to maneuver the robot to achieve the desired differentiation between potential environments. For this, we use the OpenGL specification for rendering graphics of single cubes with different color gradients to show differences visually between trajectories which an agent could take around an object and to aid with considering how a control design that takes these types of images as models might be formulated [3]. We discuss how the character of the image data should inform the design of the objective functions, modeling strategies, and constraints in a model predictive control formulation that makes predictions of future images to aid in navigation. We also discuss what is gained and lost if full image prediction is not performed when trying to sort out the environment description.

[1] Bochkovskiy, A., C.-Y. Wang, and H.-Y. M. Liao. 'Yolov4: Optimal speed and accuracy of object detection.' *arXiv preprint arXiv:2004.10934* (2020).

[2] Hutter, M., C. Gehring, A. Lauber, F. Gunther, C. D. Bellicoso, V. Tsounis, P. Frankhauser, R. Diethelm, S. Bachmann, M. Bloesch, H. Kolvenbach, M. Bjelonic, L. Isler, and K. Meyer. 'Anymal-toward legged robots for harsh environments.' *Advanced Robotics* 31.17 (2017): 918-931.

[3] Oyama, H., D. Messina, R. O'Neill, S. Cherney, M. Rahman, K. K. Rangan, G. Gjonaj, and H. Durand, 'Test Methods for Image-Based Information in Next-Generation Manufacturing,' *Proceedings of the IFAC Symposium on Dynamics and Control of Process Systems, including Biosystems (DYCOPS)*, Busan, Republic of Korea, in press.

[4] Pereira, A., and C. Thomas. 'Challenges of machine learning applied to safety-critical cyber-physical systems.' *Machine Learning and Knowledge Extraction* 2.4 (2020): 579-602.

[5] Tan, J., T. Zhang, E. Coumans, A. Iscen, Y. Bai, D. Hafner, S. Bohez, and V. Vanhoucke. 'Sim-to-real: Learning agile locomotion for quadruped robots.' *arXiv preprint arXiv:1804.10332* (2018).

[6] Tullu, A., B. Endale, A. Wondosen, and H.-Y. Hwang. 'Machine learning approach to real-time 3D path planning for autonomous navigation of unmanned aerial vehicle.' *Applied Sciences* 11.10 (2021): 4706.

[7] Yu, L., E. Yang, P. Ren, C. Luo, G. Dobie, D. Gu, and X. Yan. 'Inspection robots in oil and gas industry: a review of current solutions and future trends.' *2019 25th International Conference on Automation and Computing (ICAC)*. IEEE (2019).

222a Keynote Talk-Industrial Internet of Things at Dow Inc.

Mary Beth Seasholtz

The Operations/Manufacturing division of Dow, Inc. has been investing in Manufacturing 4.0 technologies since 2017, including those considered under the industrial internet of things (IIoT) umbrella. Data Science / Analytics is one such field. For analytics to be successful, several base technologies must be in place. (1) revealing measurements must be made to interrogate the process at appropriate locations and (2) powerful analytics models typically require data from various sources. Therefore, the assembly of varied data must be possible. Scenarios exist where pertinent data are collected by a third party; for example, a provider of services such as water treatment or equipment maintenance. These data need to be automatically sent from the third party into Dow's data systems and ultimately fed into the analytics model. (3) The final base component in a successful analytics application is clear understanding of the user experience. We must be able to help the user realized what they will learn from the analytics, what they will do with that learning, how will the encounter with the model make their job easier, more efficient.

222c Sensor Placement for Process Networks Based on the Sensitivity Analysis

Siyu Liu, Jinfeng Liu, Xunyuan Yin

Sensor placement or selection is one fundamental problem in process control and arises in various applications including target tracking, agro-hydrological systems, power systems, and wireless networks. Sensor placement has been a computationally challenging problem when the number of potential sensors is large. In this work, we propose a sensitivity based approach to perform sensor placement or selection and the proposed approach is much more computationally efficient than the existing approaches.

In sensor placement, observability plays an important role. The observability Gramian for linear system or the empirical observability Gramian has been often used in sensor placement. Different approaches based on the observability Gramian have been developed such as approach based on maximizing the trace of the Gramian [1], maximizing the determinant of the Gramian [2]. When using the trace of an observability Gramian matrix as the measure of observability, it in many cases may lead to meaningless results since a system can be unobservable if it has an eigenvalue of zero, despite a potentially large trace. When the determinant of a Gramian matrix is used as a measure of the observability, it is possible to use the smallest eigenvalue as an additional measure in order to avoid the situation that the eigenvalue of the Gramian matrix is almost zero and the system is not fully observable. However, one major limitation of the above methods is that the computational complexity is very high when applied to nonlinear systems, because these methods would require simulating the system many times for each perturbed initial condition in order to compute the ensemble average.

Graph-based methods are another class of popular methods used for the sensor selection in networked systems based on the structural observability theory [3]. By examining the strongly connected components (SCCs) in a graph describing node connections, graph-based methods can determine the number and locations of sensor nodes for ensuring structural observability [4,5]. The minimal set includes one sensor in each root SCC (an SCC with no incoming edges) of the observability inference diagram. Graph-based methods has become an influential method and has been used to provide insights into the relation between network topology and observability. However, since these approaches do not explicitly take into account model parameters, in some cases, the estimator designed based on the sensors selected by the graph-based approaches may lead to less accurate state estimates [6]. A workaround is to simultaneously deal with the sensor selection and state estimation, which has the potential to improve the overall estimation performance [7].

Based on the above observations, in this work, we propose a sensitivity-based approach to determine the minimum number of sensors and their optimal locations for state estimation. The proposed approach is much more computationally efficient compared with existing methods. In the proposed approach, the local sensitivity matrix of the measured outputs to initial states is used as a measure of the degree of observability. The minimum number of sensors is determined in a way such that the local sensitivity matrix is full column rank. The subset of sensors that satisfies the full-rank condition and provides the maximum degree of observability is considered as the optimal sensor placement. Successive orthogonalization of the sensitivity matrix is conducted in the proposed approach to significantly reduce the computational complexity in selecting the sensors. To validate the effectiveness of the proposed method, it is applied to two processes including a chemical process consisting of four continuous stirred-tank reactors and a wastewater treatment plant. In both cases, the proposed approach can obtain the optimal sensor subsets.

References

[1] A. K. Singh, J. Hahn. Sensor location for stable nonlinear dynamic systems: Multiple sensor case. Industrial andamp; Engineering Chemistry Research, 45(10):3615-3623, 2006.
[2] J. Qi, K. Sun, W. Kang. Optimal PMU placement for power system dynamic state estimation by using empirical observability Gramian. IEEE Transactions on Power Systems, 30(4):2041-2054, 2015.
[3] J. D. Stigter, D. Joubert, J. Molenaar. Observability of complex systems: Finding the gap. Scientific Reports, 7(1):16566, 2017.
[4] Y.-Y. Liu, J.-J. Slotine, A.-L. Barabasi. Observability of complex systems. Proceedings of the National Academy of Sciences, 110(7):2460-2465, 2013.
[5] M. T. Angulo, A. Aparicio, C. H. Moog. Structural accessibility and structural observability of nonlinear networked systems. IEEE Transactions on Network Science and Engineering, 7(3):1656-1666, 2019.
[6] A. Haber, F. Molnar, A. E. Motter. State observation and sensor selection for nonlinear Networks. IEEE Transactions on Control of Network Systems, 5(2):694-708, 2017.
[7] A. Haber. Joint sensor node selection and state estimation for nonlinear networks and systems. IEEE Transactions on Network Science and Engineering, 8(2):1722-1732, 2021.

222d Development of AI Algorithms for Corrosion Prediction in Midstream Industry

Helen Lou, Jian Fang, Sidney Lin

Corrosion has been causing the mid-stream industry billions of dollars per year. It is highly desirable to predict pipeline corrosion. For natural gas pipelines, the species cause most concerns on corrosion are water, hydrogen sulfide, carbon dioxide, and oxygen. In addition, the temperature, pressure, and velocity also make impacts. For crude oil pipelines, the species are more complicated.

In this research, the corrosion rate data was generated from sound first-principles based models. Then accurate AI algorithms were developed to predict the corrosion rates in natural gas and crude oil pipelines. This research will help bridge the gap from "post-corrosion" measurement to proactive corrosion management, revolutionize the practice of corrosion management, enhance the reliability of mid-stream assets, improve operation efficiency, and reduce potential environmental risks.

222f Physics-Informed Surrogate Models for Manufacturing Applications

Utsav Awasthi, George M. Bollas

Effective monitoring of a physical or a chemical process requires some form of a representation (inference, clustering, classification) of the measured signals of the process. In model-based monitoring methods, first principles or data-driven models are used to calculate deviations from what is anticipated for the process and what is measured from the available sensors. First-principles models accurately capture the process behavior, but their development can be cumbersome, as physics might be unknown, or the exact relationship or correlation of process inputs and outputs is not accurate or does not exist. Recent advancements in Industry 4.0 technology have led to rich data sets to assist with the development of machine learning data-driven models for process monitoring. These data-driven models can be made more robust and intelligent by incorporating process knowledge (Sansana et al., 2021). The models obtained by fusing process knowledge with data-driven models are novel forms of hybrid modeling. In this context, surrogate models can be seen as a type of hybrid models that are transparent, simple to execute, and wherein domain expert knowledge can be incorporate in the form of intelligent basis functions. In turn, these domain-expert-informed models can be leveraged to related process signals or features with process health monitoring.

In this work, we show the development of a physics-informed surrogate model and its use as an inferential sensor to perform fault classification for process monitoring. Surrogate models are approximate models in the form of explainable algebraic expressions in terms of the inputs of the system. The model outputs are the desired process information or signals that cannot be measured directly, for which it is difficult to have accurate first-principles models. The proposed surrogate models were developed with data of system inputs and outputs for precision machining processes. The process knowledge was incorporated in the form of custom basis functions. We evolved surrogate models using two methodologies: the Automated Learning of Algebraic Models for Optimization (ALAMO) (Cozad et al., 2014; Cozad et al., 2015; Wilson and Sahinidis, 2017); and GPTIPS (Searson, 2015). ALAMO uses machine learning, statistical, and optimization methods to develop a surrogate model, whereas GPTIPS utilizes multigene genetic programming-based symbolic regression to generate models based on a Pareto optimal solution between the model complexity and accuracy. Both methodologies are forms of symbolic regression of the system outputs with respect of its inputs. The models developed were explored for their applicability in classification of faults using K-nearest neighbors.

The process monitoring approach mentioned above was applied to a manufacturing process. Tool wear in machining is the leading cause of tool failure, machine failure, and improper surface finish, resulting in downtime and energy waste in manufacturing. Tool wear depends on machine settings, material and tool properties. Generic surrogate models for tool wear were developed using ALAMO and GPTIPS for data collected from the open literature and in-house run-to-failure tests. Inputs to the surrogate models were the cutting and machine settings, as shown in Fig. 1. Process knowledge was incorporated in the form of custom basis functions, which were formed using literature reports on the mechanisms of tool wear and the key factors contributing to it. These models were used as inferential (soft) sensors for tool wear classification between a new tool and a tool with a flank wear of 0.2 mm, which is commonly a threshold for tool replacement. These soft sensors were shown to improve the classification of tool wear compared to just using the hard sensors of the machine. Further, the surrogate model can be used to improve the prediction of model outputs of the machining model proposed in Awasthi et al. (2021) by providing a generic expression of tool wear propagation. Therefore, utilizing the physics-informed tool wear surrogate model helps increase the effectiveness of the tool wear detection process to avoid downtime and energy waste.

Acknowledgments:

This material is based upon work supported by the U.S. Department of Energy's Office of Energy Efficiency and Renewable Energy (EERE) under the Advanced Manufacturing Office Award Number DE-

EE0007613. We also gratefully acknowledge the Air Force Research Laboratory, Materials and Manufacturing Directorate (AFRL/RXMS) for support via Contract No. FA8650-20-C-5206.

Disclaimer: This report was prepared as an account of work sponsored by an agency of the United States Government. Neither the United States Government nor any agency thereof, nor any of their employees, makes any warranty, express or implied, or assumes any legal liability or responsibility for the accuracy, completeness, or usefulness of any information, apparatus, product, or process disclosed, or represents that its use would not infringe privately owned rights. Reference herein to any specific commercial product, process, or service by trade name, trademark, manufacturer, or otherwise does not necessarily constitute or imply its endorsement, recommendation, or favoring by the United States Government or any agency thereof. The views and opinions of authors expressed herein do not necessarily state or reflect those of the United States Government or any agency thereof."

References

Sansana, J., Joswiak, M. N., Castillo, I., Wang, Z., Rendall, R., Chiang, L. H., andamp; Reis, M. S. (2021). Recent trends on hybrid modeling for Industry 4.0. Computers and Chemical Engineering, 151, 107365. https://doi.org/10.1016/j.compchemeng.2021.107365

Cozad, Alison, Nikolaos V. Sahinidis and David C. Miller. "Learning surrogate models for simulation-based optimization." AIChE Journal 60 (2014): 2211-2227.

Cozad, A., Sahinidis, N. V., andamp; Miller, D. C. (2015). A combined first-principles and data-driven approach to model building. Computers and Chemical Engineering, 73, 116–127. https://doi.org/10.1016/j.compchemeng.2014.11.010

Wilson, Z. T., andamp; Sahinidis, N. V. (2017). The ALAMO approach to machine learning. Computers and Chemical Engineering, 106, 785–795. https://doi.org/10.1016/j.compchemeng.2017.02.010

Searson, D. P. (2015). GPTIPS 2: An Open-Source Software Platform for Symbolic DataMining. In A. H. Gandomi, A. H. Alavi, andamp; C. Ryan (Eds.), Handbook of Genetic Programming Applications (pp. 551–573). Cham: Springer International Publishing. https://doi.org/10.1007/978-3-319-20883-1_22

Awasthi, U., Wang, Z., Mannan, N., Pattipati, K., andamp; Bollas, G. M. (2021). Physics-based Modeling and information-theoretic sensor and settings selection for tool wear detection in precision machining. In Process.

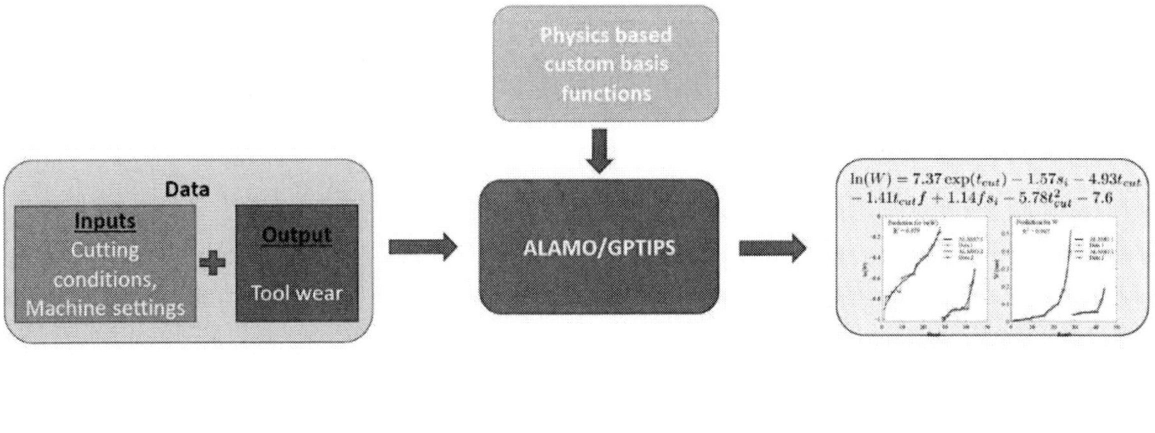

222g Enhancing Operator's Trust in AI-Based Process Monitoring Technologies: Providing Explanations for Multi-Mode Processes

Abhijit Bhakte, Piyush Kumar Kumawat, Rajagopalan Srinivasan

In process industries, effective handling of abnormal events entails detecting and diagnosing issues promptly, as well as providing fast and reliable decision support to operators so they may take appropriate steps to avoid disturbances from escalating. With the advent of Industry 4.0, industrial processes now have unique opportunities to use artificial intelligence methods to detect and correct operational abnormalities. Deep learning methods for Fault Diagnosis and Detection (FDD) have received significant attention in the last few years[1]. However, the inability of black-box deep learning systems to provide human-interpretable outputs is a significant shortcoming that prevents its widespread adoption. Plant operators need intelligent systems that correctly identify faults and provide explanations for the results generated by Deep Neural Networks (DNNs). These explanations will enhance the plant operator's trust in the DNN by providing the reasons for the DNN's prediction. Using these, the operator can independently verify if they can trust the predictions and hence use the DNN with confidence. We propose such a system in this work.

Explainable Artificial Intelligence (XAI) is an emerging field of research that aims at explaining predictions of DNNs. It seeks to interpret the predictions that humans can understand by assigning a relevance or contribution to each input variable for a given sample. In XAI research, two separate approaches have been pursued, i.e., inherently interpretable AI methods (such as decision trees) and posthoc methods [2]. Posthoc approaches are used to explain pre-developed models; they regard the model as a black box and have no effect on the model's performance. Recently, XAI-based strategies have successfully explained the results generated by DNNs for FDD using different methods, including Integrated Gradients [3] and Layerwise Relevance Propagation (LRP)[4].

Previously [3], we presented a method for the interpretability of DNNs used for fault diagnosis using the Integrated Gradients (IG) method. It is motivated by the concept of Shapely value [5] derived from cooperative game theory. The IG is a gradient-based attribution, a technique to explain DNNs by attributing them to the neural network's inputs. The obtained attributions are the summation of the difference between the target output and the output as evaluated at the baseline [6]. These are used during real-time process monitoring to identify key variables responsible for the fault. Explanations are calculated using a windowed attribution scheme to make them robust to process noise. An inseparability metric is also used to highlight the most pertinent explanatory variables.

The above IG-based method and other similar explainability techniques reported in the literature are limited to the situation where the process has one nominal operating region. These assumptions often do not hold for real industrial chemical processes, where several operating modes may be prevalent even during nominal operations [7]. In such cases, the previously developed XAI methods fail to provide correct explanations even when the process is operating under normal conditions since any deviation from the baseline is considered to be a change. We seek to overcome this limitation in the current work.

In this work, we report a novel strategy for generating explanations of a DNN developed for multi-mode operation. The proposed methodology comprises two distinct components – (a) a bank of XAI models, and (b) a real-time model selector. The bank of IG-based XAI models, one for each operating mode, is developed offline. Each of these XAI models relies on data and DNN results from an operating mode and uses a mode-specific baseline. Hence it can provide accurate explanations –

attributions to the input process variables - when the plant operates in the corresponding mode. The model-selector component utilizes real-time process data to identify the plant's operating mode dynamically and hence determines the XAI model that should be used to obtain explanations [8]. The efficiency of the proposed methodology is demonstrated using the Tennessee-Eastman challenge process [9]. In this paper, we will describe the overall architecture of the proposed technique. A comparison of the results from the multi-mode IG with those from a traditional IG will also be reported.

References

[1] Z. Jiao, P. Hu, H. Xu, and Q. Wang, "Machine Learning and Deep Learning in Chemical Health and Safety: A Systematic Review of Techniques and Applications," *ACS Chem. Health Saf.*, vol. 27, no. 6, pp. 316–334, 2020, doi: 10.1021/acs.chas.0c00075.

[2] F. K. Dosilovic, M. Brcic, and N. Hlupic, "Explainable artificial intelligence: A survey," in *41st International Convention on Information and Communication Technology, Electronics and Microelectronics (MIPRO)*, 2018, pp. 210–215. doi: 10.23919/MIPRO.2018.8400040.

[3] A. Bhakte, V. Pakkiriswamy, and R. Srinivasan, "An Explainable Artificial Intelligence Based Approach for Interpretation of Fault Classification Results from Deep Neural Networks," *Chem. Eng. Sci.*, p. 117373, 2021, doi: https://doi.org/10.1016/j.ces.2021.117373.

[4] P. Agarwal, M. Tamer, and H. Budman, "Explainability: Relevance based dynamic deep learning algorithm for fault detection and diagnosis in chemical processes," *Comput. Chem. Eng.*, vol. 154, p. 107467, 2021, doi: https://doi.org/10.1016/j.compchemeng.2021.107467.

[5] L. S. Shapley, "17. A Value for n-Person Games," in *Contributions to the Theory of Games (AM-28), Volume II*, Princeton University Press, 2016, pp. 307–318. doi: doi:10.1515/9781400881970-018.

[6] M. Sundararajan, A. Taly, and Q. Yan, "Axiomatic attribution for deep networks," in *34th International Conference on Machine Learning, ICML 2017*, 2017, vol. 7, pp. 5109–5118. [Online]. Available: arXiv:1703.01365

[7] S. J. Zhao, J. Zhang, and Y. M. Xu, "Monitoring of Processes with Multiple Operating Modes through Multiple Principle Component Analysis Models," *Ind. Eng. Chem. Res.*, vol. 43, no. 22, pp. 7025–7035, Oct. 2004, doi: 10.1021/ie0497893.

[8] R. Srinivasan, P. Viswanathan, H. Vedam, and A. Nochur, "A framework for managing transitions in chemical plants," *Comput. Chem. Eng.*, vol. 29, no. 2, pp. 305–322, 2005, doi: https://doi.org/10.1016/j.compchemeng.2004.09.024.

[9] J. J. Downs and E. F. Vogel, "A plant-wide industrial process control problem," *Comput. Chem. Eng.*, vol. 17, no. 3, pp. 245–255, Mar. 1993, doi: 10.1016/0098-1354(93)80018-I.

222h Evaluating Controllers for Next-Generation Manufacturing

Fnu Akkarakaran Francis Leonard, Keshav Kasturi Rangan, Ilham Azali Assoumani, Jacob Noll, Carley Fields, Nazir Jairazbhoy, Emmanuel Dannug, James Redman, Helen Durand, K.Y. Simon Ng

The next generation of manufacturing for the chemical and materials industries involves a variety of new considerations for which it will be necessary to develop good strategies for evaluating how controllers behave. For example, with cybersecurity of control systems as an important concern [1] with increasing emphasis on data and automation, it is important to consider how process design and control interact in making processes more resilient to attacks. Furthermore, the next generation of materials considerations may involve not only new control methods that might, for example, take advantage of image-based control, but also potentially materials that are themselves controlled. Developing techniques for evaluating these various types of considerations computationally to reduce testing efforts while verifying safe and on-specification operation is important.

This talk will review a series of case studies where we explore ideas for evaluating control for next-generation manufacturing processes as diverse as geothermal energy production, self-assembly with an assumed image-based sensor, and materials that respond to changes in their environment. In the first part of the talk, we discuss how steady-state simulation studies of a geothermal energy process can guide an understanding of how different designs for a geothermal energy system contribute to the severity of different types of cyberattacks on the control systems in the sense of how much they impact the production of electricity [2]. We discuss how dynamic simulation might contribute to further analysis. We provide thoughts on how attacks seeking to impact manufacturing could also have impact by modifying peoples' behavior. We close by presenting a perspective on how next-generation materials design and manufacturing can be impacted by new concepts for testing and understanding control for such systems. For example, inspired by the concept that we might use the software Blender for image-based controller evaluation [3,4], we discuss how more sophisticated examples of image-based control than were previously presented in [3,4] might be considered. For example, inspired by cases where image processing has been integrated with self-assembly control [5], we discuss preliminary steps that would need to be taken if one wants to build an image-based control simulation for such a process in Blender to test the effectiveness of various image-based control laws without an experimental unit. Finally, we present thoughts related to the control of materials, particularly how control theory intersects with modeling strategies such as molecular dynamics. While control for molecules at the quantum level has been of interest [6], we discuss molecular dynamics simulation [7] due to its closer framework to the traditional ordinary differential equations models used in control theory for chemical processes.

[1] McLaughlin, S., Konstantinou, C., Wang, X., Davi, L., Sadeghi, A. R., Maniatakos, M., andamp; Karri, R. (2016). The cybersecurity landscape in industrial control systems. Proceedings of the IEEE, 104(5), 1039-1057.

[2] Rangan, K. K., J. Abou Halloun, H. Oyama, I. Azali Assoumani, N. Jairazbhoy, H. Durand, and S. K. Ng, "Design and Control Resilience/Robustness: Relationships to Quantum Computing and Cybersecurity," Proceedings of the IFAC Symposium on Dynamics and Control of Process Systems, including Biosystems (DYCOPS), Busan, Republic of Korea, in press.

[3] Oyama, H., D. Messina, R. O'Neill, S. Cherney, M. Rahman, K. K. Rangan, G. Gjonaj, and H. Durand, "Test Methods for Image-Based Information in Next-Generation Manufacturing,"

Proceedings of the IFAC Symposium on Dynamics and Control of Process Systems, including Biosystems (DYCOPS), Busan, Republic of Korea, in press.

[4] Oyama, H., A. F. Leonard, M. Rahman, G. Gjonaj, M. Williamson, and H. Durand, "On-line Process Physics Tests via Lyapunov-based Economic Model Predictive Control and Simulation-Based Testing of Image-Based Process Control," Proceedings of the American Control Conference, Atlanta, Georgia, in press.

[5] Grover, M. A., Griffin, D. J., Tang, X., Kim, Y., andamp; Rousseau, R. W. (2020). Optimal feedback control of batch self-assembly processes using dynamic programming. Journal of Process Control, 88, 32-42.

[6] Magann, A. B., Grace, M. D., Rabitz, H. A., andamp; Sarovar, M. (2021). Digital quantum simulation of molecular dynamics and control. Physical Review Research, 3(2), 023165.

[7] Rapaport, D. C., andamp; Rapaport, D. C. R. (2004). The Art of Molecular Dynamics Simulation. Cambridge University Press.

39a Using Digital Technologies to Optimize Plant Uptime

Jonas Norinder

This presentation discusses how recent advances in digital technologies can be leveraged to optimize maintenance activities to optimize plant uptime. This will be illustrated through several specific use cases that focus on how mobile devices such as tablets can be linked remotely to monitoring and control facilities from multiple locations. Maintenance crews can also collaborate in real time with remote experts—through audio, video, or chat—to improve asset management and maintenance planning. Augmented Reality advances enhance the experience, allowing the remote expert, for instance, to overlay repair instructions onto the shared image. Smart apps help integrate tablets into the plant environment, enabling access to asset-specific information, including past maintenance and recent operations data.

Digitization solutions such as these also draw on advances of the digital twin concept. Because a wealth of equipment and operations data is available digitally today, it can be downloaded to the maintenance person's tablet for immediate use. This gives maintenance workers everything they need where and when they need it to resolve equipment problems.

During this presentation you will gain the following insights:

- A general understanding of key digitalization topics such as the digital twin, 3D, simulation as well as innovative ways of collecting data
- Learn how information and data can be used to operate and maintain a plant more efficiently
- Understand how maintenance can move from being reactive to predictive

39b The Necessity of a Digital Toolbox

Wolter Last

Introduction

As the Covid-19 pandemic evolves into an endemic, companies are returning to work under "new normal" conditions. Some employees are back onsite, while others are working in a hybrid mode (onsite only part-time), and still others are working solely from a home office. More than ever before, this calls for remotely operated and controlled asset management systems to keep plant machinery running smoothly. In particular, the pandemic jolted the oil, gas and petrochemical industry into automating its processes and doing more things remotely, requiring fewer personnel.

The days of Excel spreadsheets and disparate, standalone systems that address part of the plant's operations are definitely over. Team members now need to be in sync, sharing information across multiple locations and disciplines for effective collaboration and critical decision-making. Recent advancements in emerging technologies, like cloud computing, have helped to make the Digital Age a reality, with anytime, anywhere access to real-time data and safeguards to keep that data secure. This has paved the way for a comprehensive tool, or set of integrated tools, to combine engineering and maintenance tasks into a centralized application for optimal asset management.

Challenges

Cloud-based asset management software is a new way of working within the industry. The cloud is a server cluster that can dynamically allocate tasks. If one of the servers fails to work, the other servers automatically take over, and they will continue as nothing happened. On-site, it is done through redundancy and virtual machines, but this needs to be supported by the IT team.

Nowadays, cloud service providers can outsource these operations, such as Microsoft Azure. The cloud provider handles complex topics like cyber security and disaster recovery, which often cause problems in the organization.

Moving from an on-site to an online cloud-based system in a secure way also provides more accessibility for the employees. The information which only was available in the office is now accessible from home or mobile. But accessibility is not the only benefit of using a cloud system. Comparing it to a traditional on-premises system, there are a few significant advantages.

3 Advantages

Faster setup

It is a matter of minutes to set up a new environment with a cloud-based system. The difference here with an on promise virtual machine comes in the flexibility of resources. When a virtual environment on-premise runs out of space or becomes slow, new parts or entire new servers need to be ordered to resolve the issue. When more resources are required for a cloud environment, the cloud provider makes sure that spare capacity is monitored and available when needed. Applications that run in the cloud do not need client-side installation, making a start and maintenance of a project more accessible and quicker.

Lower IT-related costs

Cloud environments provide better up times through the dynamic allocation of the servers it runs on. If an on-premises server has a CPU failure, that server stops working and needs a replacement before the applications that are run on the server can be used again. With cloud environments, another server within the cloud will dynamically take over the tasks of the failing server.

Cloud environments managed by cloud providers have the added benefits of requiring less local IT resources and knowledge. Items like cyber security and server management are regular tasks for a cloud provider, and this might not be the case for a company with its own IT infrastructure. Moving from a local closed-off network towards an internet-enabled network brings a new list of challenges that need to be handled with great care.

Accessibility andamp; Security

The door to remote access can be controlled and opened with centralized cyber security management. This means that information that was previously only available at the office is now useable from home or other locations which it previously was not. This can also provide suppliers with information to help fine-tune performance issues.

Maintainability

It is important to keep software up to date. Wishes from customers change, developers develop ideas develop, and third parties' requirements increase. To satisfy all these changes, software needs to be continuously evolving. In a closed-off environment, this is challenging as the supplier cannot quickly provide updates. Supporting and troubleshooting is also a costly effort.

24/7 Monitoring

If the server goes down with an on-premises environment, local resources need to find out and get it running again. If the server goes down on the weekend, the software could be down for days at a time. Cloud-hosted systems reside on servers that are monitored around the clock. The servers are typically co-located in facilities that house other applications to spread monitoring costs across the entire customer base. Resources are available to ensure that applications keep running or are recovered quickly.

A digital toolbox

Manual and de-centralized data entry into disconnected systems needs to be replaced by a single "digital" toolbox, enabling asset management to be accomplished via analyzers more effectively and more efficiently. One, all-encompassing application—from analysis to maintenance—would allow technicians, engineers, operators, and managers to see everything and, importantly, to be on the same page for increased collaboration. With digital connectivity, this concept delivers information about equipment condition/health in real time for swift, proactive maintenance to be performed, thus, extending equipment life. A centralized, web service can be hosted from on-premises or via the cloud, depending on the choice of network architecture. Remote access can be provided by any available means, such as a remote-control center, to conduct equipment monitoring and control activities. The remote-control center also provides mobile access possibilities to the toolbox which provide on hand information and logging capabilities

What needs to be in the toolbox?

The key factor is consolidation! The following basic items comprise a consolidated digital toolbox for oil and gas operators:

Live data from the field

It all starts with needing to know the "live" status of the equipment in the field, especially when working at a distance (e.g. during Covid times). This includes process information, diagnostics, validation data, maintenance records, etc. Field operators need to look at data in its actual state, and then be able to escalate any problems that might require further analysis, maintenance, or even shutdown.

Tools to help find problems

Statistical process control mechanisms build calculations that trigger events and actions (alarms, emails, reports, validations, etc.), automate knowledge of SMEs to find root cause problems and define trigger parameters, such as KPIs, and track performance in time for historical reference and trending.

However, the automated tool cannot do it all; human intervention is needed to set the proper limits and to fully understand the root cause of a problem that might require onsite action by a trained technician. Prediction is as powerful as the person making the algorithm for the prediction.

Maintenance scheduling andamp; tracking

Live data from the field opens opportunities to create maintenance tasks in a timely manner based on pre-defined triggers for desired levels, e.g. if downhole temperature is too low or too high, if pressure drops, or if fluid flow rate slows (indicating a filter may be plugged/needs replacement).

Reporting

reports to designated individuals. This alleviates the tedious task of creating the same routine reports, allowing users to schedule when a report is needed, what report template to use, and then automatically email it to management. The time it takes to manually set up each report can be put to better use by focusing on value-added work for the company.

Summary

The digital toolbox has proved itself in the field, as described above. The cast of tools is already there, but more tools have yet to be explored and added as this technology approach matures. Regardless of technological advancements, however, it will always tie back to subject matter experts putting their know-how into the tools, so it can become the best set of tools for technicians in the field. This involves a transfer of knowledge from boomers to the next generation—preserving accumulated experience from over the years in the toolbox!

While a plethora of services and software companies (e.g. SAP, Maximo, PI, and Hint, to name a few) are working in the same direction, they do not necessarily share the same vision. The recommended route is to ideally have everything in one digital toolbox and make it available as broadly as possible (web-accessible/networked) to support remote workers 24/7. It also has to be able to escalate issues to supervisors/managers in order to get technical help from experts and suppliers, if needed. The AML Information Management toolbox from Hint Global is a viable solution, ready to go.

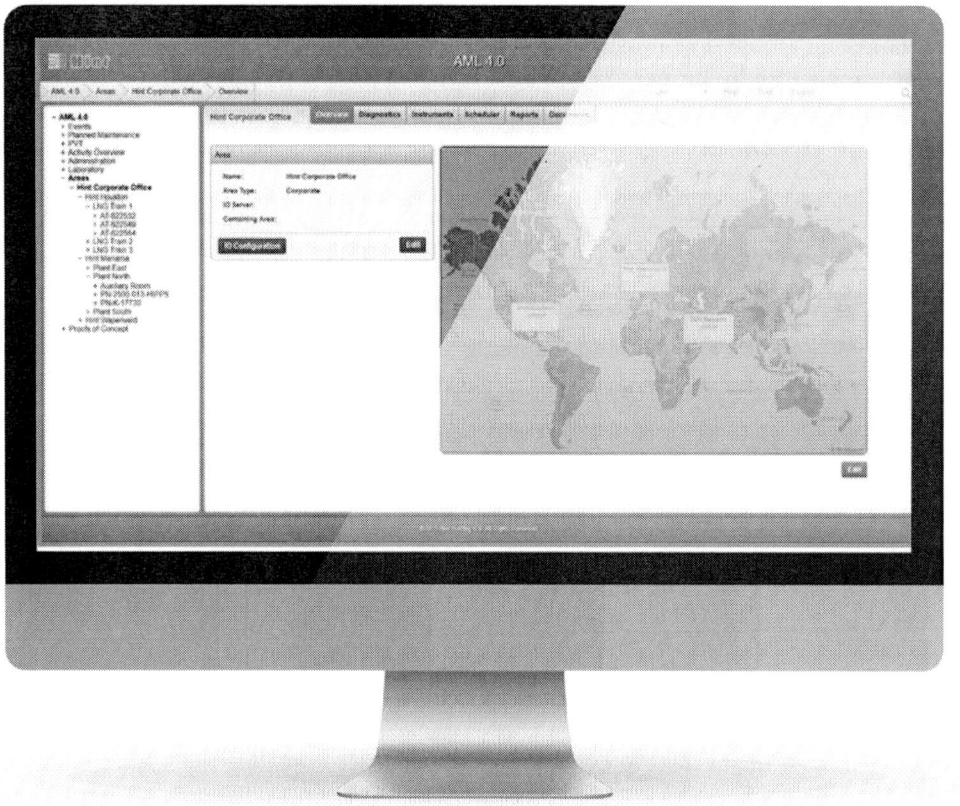
AML 4.0
AML 4.0
Areas
Hint Corporate Office
Overview
Hint Corporate Office
Diagnostics
Instruments
Scheduler
Reports
Events
Planned Maintenance
PVT
Activity Overview
Administration
Laboratory
Areas
Hint Corporate Office
Hint Houston
LNG Train 1
LNG Train 2
LNG Train 3
Plant East
Plant North
Auxiliary Room
Plant South
Proofs of Concept
Area
Name:
Hint Corporate Office
Area Type:
Corporate
IO Server:
Containing Area:
IO Configuration
Edit
Edit

39c Business Intelligence: A Case Study of a Chemical Plant

Abner Colman, Natalya A. B. Almeida, Fernanda Andrade, Heloyse Reges, Fernando V. Lima, Heleno Bispo

Data management has been a significant challenge for companies in market competitiveness. The 4th Industrial Revolution has transformed the industry into a strategic field based on the integration of technologies through which it is possible to fulfill the needs of large companies[1,2]. In the chemical industry, the so-called intelligent industry is already executing a set of technologies based on the interaction between the process and the data analysis systems. A particular data analysis system is known as Business Intelligence (BI), which is used as a Decision Support System (DSS)[2]. It consists of collecting, monitoring, sharing, and organizing data, with the aim of obtaining a good set of performance indicators generating, in this way, useful information that manages the business and supports the decision-makers[3,4].

In this sense, this work proposes the generation of a Digital Shadow and implementation of a complete BI system from data generated by a rigorous Process Digital Twin[5]. For this purpose, a chemical process has been implemented in Aspen Plus Dynamics (Aspentech), where data has been continuously generated and connected by an OPC Matrikon Architecture. Such data has been collected by a PI System connector (PI ICU – Interface Configuration Utility) and then sent to a PI System Architecture (AVEVA/OSIsoft). With this software infrastructure, it is possible to capture, store, organize, process, analyze, and share the information obtained/treated through the acquired data[6].

As a result, a complete system has been developed in which it's possible to monitor KPI's on the web in real time (Figure 1A) by PI Vision (AVEVA/OSIsoft). In this environment, as depicted in Figure 2B, it is possible:

- Establishing data access according to organizational hierarchies;

- Predicting conditions based on data history;

- Obtaining valuable information based on KPI's to be used by decision-makers, such as:

    - Equipment: Status, Active Machines, Availability, Performance, Quality, and OEE;

    - Production: Revenue, Percentage of main components, Production rate, and Utilities;

    - Events: Section Downtime; Controller Downtime and Cashflow.

- Generating an automatic report system based on event frames and notifications by email.

Finally, it has become clear that industries that use and reuse data most efficiently will have a superior competitive advantage, and this approach can be accomplished through the implementation of the BI system in which it is feasible to identify new possibilities and strategies based on data.

References

[1] SCHWAB, Klaus. The Fourth Industrial Revolution. January 2016, World Economic Forum. ISBN 1944835008.

[2] AIN, Noor Ul; VAIA, Giovanni; DELONE, William H.; WAHEED, Mehwish. Two decades of research on business intelligence system adoption, utilization and success – A systematic literature review. Decision Support Systems. Vol. 125, ISSN 0167-9236. 2019.

[3] ŽIŽEK, S. S.; NEDELKO, Z.; MULEJ, M.; ČIČ, Ž. V. Key Performance Indicators and Industry 4.0 –

A Socially Responsible Perspective. Naše gospodarstvo / Our Economy, 66(3), p. 22–35. 2020. DOI: 10.2478/ngoe-2020-0015.

[4] ZORZUT, Sebastjan. Vodenje proizvodnje v procesni industriji z upoštevanjem ključnih kazalnikov učinkovitosti : doktorska disertacija. Univerza v Ljubljani, Fakulteta za elektrotehniko. 2009.

[5] KUEHN, Wolfgang. Digital twins for decision making in complex production and logistic enterprises. International Journal of Design andamp; Nature and Ecodynamics, Vol. 13, No. 3, pp. 260-271. 2018.

[6] BASCUR, Osvaldo A.; O'ROURKE, Jim. Chapter 6 - Measuring, managing, and transforming data for operational insights. Smart Manufacturing. pg 129-165. Elsevier. 2020.

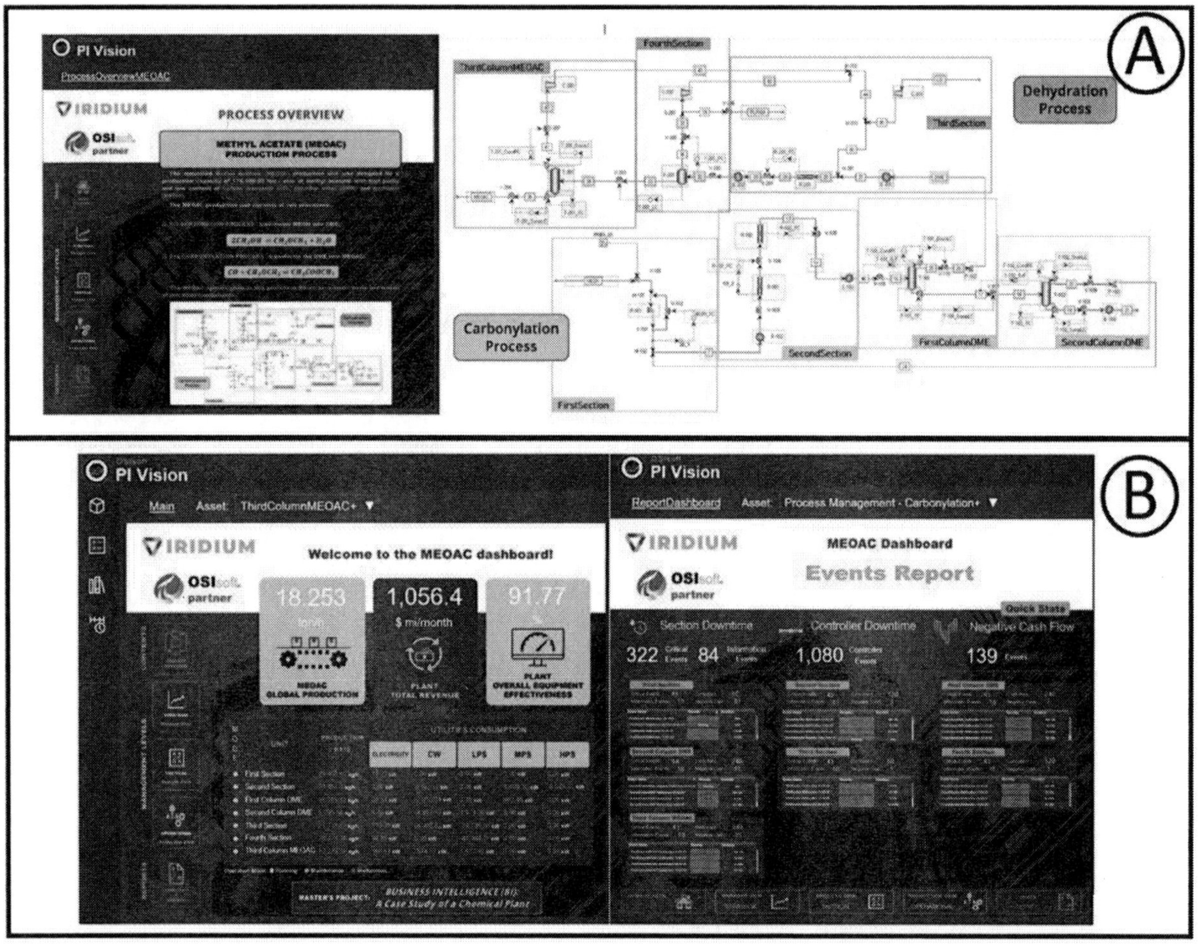

39d Improving Control Room Operator Performance during Training Using a Cognitive Digital Twin

Bharatwaajan B, Babji Srinivasan, Rajagopalan Srinivasan

Abstract

Process industries employ control operators to oversee plant operations. With the presence of advanced state of the art automation, the role of an operator in a process industry has shifted towards monitoring the plant and making decisions during abnormalities (Kluge et al., 2015). As operators are increasingly being expected to deal with abnormal situations, operator training and learning becomes significant. It is a common practice in the industry to use Operator Training Simulators (OTS) for training operators. Modern training simulators log the actions taken by an operator in response to situations to give feedback and understand operator's learning of the process (Manca et al., 2014). Traditionally, operator performance in these simulators is studied by analysing the operator actions such as time to respond, number of actions taken, accuracy of the actions etc. (Adhitya et al., 2014). However, these methods do not account for the cognitive behaviour of operators. Cognitive aspects such as cognitive workload are important to assess operator performance during training (Huey and Wickens, 1993); increase in cognitive workload results in degradation of performance. Our previous research using eye-tracking reveals exciting insights into the cognitive behavior of control room operators. For instance, we found that there is a difference in gaze pattern of operators who successfully controlled the plant during abnormality against operators who failed to complete the task. Eye tracking is based on the "eye-mind" philosophy (Just and Carpenter, 1976); eye-gaze provides a trace of a person's attention allocation and thus provides insights into their information acquisition pattern. In this work, we seek to utilize the difference in eye gaze pattern of experts and novices for operator training. Specifically, we propose a model-based approach using ACT-R (Adaptive Control of Thought – Rational) to obtain the eye gaze pattern of experts which is used to train the novice control room operators.

ACT-R is a theory about the working of human cognition and also provides a computational framework to model any cognitive tasks based on psychology experiments and facts. ACT-R has a symbolic component (responsible for procedural learning, decision-making) and a sub-symbolic learning (responsible for cognitive processes like thinking, learning). ACT-R also provides the locations of different modules responsible for different tasks (like decision-making, thinking, long-term memory, short-term memory, vision) inside the brain using fMRI studies (Anderson, 2009). ACT-R has been used for modelling behaviour in various safety critical domains such as aviation (Chen et al., 2021), and driving (Deng et al., 2019). In this work, we developed an ACT-R model to capture the eye gaze and actions of control room operators during plant abnormalities. The proposed ACT-R model has a total 48 production rules that modify the different modules so that it can interact with an HMI, take control actions, and learn about the process. For instance, changes in the production rule in visual module is utilized to search for various control valves in the HMI. The vision module of ACT-R also provides information about the eye gaze during the task. The eye gaze extracted from the proposed model is used to train the novice control room operators.

For development of the proposed ACT-R model, we created an ethanol production plant simulator and conducted experiments requiring a participant to control disturbances in the process. Six different scenarios were designed to test the understanding of the process. Participants would interact with the simulation using a custom designed HMI to control disturbance(s) corresponding to each scenario. Overall, ten participants performed the experiments consisting of six to ten trials for each participant with repetitive tasks. During the experiment eye gaze data, EEG, process data, and user actions data, and were recorded. We observed an improvement in the performance of participants across trials (due to learning from repetitive tasks). This is reflected in a reduced process error (measured as integral absolute abnormality), completion time, and cognitive workload across trials (Iqbal et al., 2021). This data was used to tune the production rules and parameters of the developed ACT-R model. The model learns the ethanol production process through its interaction with the HMI. It creates a mental model of the process and retrieves it to solve errors in subsequent trials. The model also replicates human subject results in a reduced process error (measured in integral absolute abnormality), completion time, and cognitive workload across trials. In this work, we will demonstrate the use of the eye gaze obtained from the proposed ACT-R model to train the control room operators during plan abnormalities.

Key words: Operator training, eye gaze, ACT-R, digital twin, Cognitive model

Acknowledgement: This work is partially funded by American Express Lab for Data Analytics, Risk and Technology, Indian Institute of Technology Madras

References

Adhitya, A., Cheng, S.F., Lee, Z., Srinivasan, R., 2014. Quantifying the effectiveness of an alarm management system through human factors studies. Computers andamp; Chemical Engineering 67, 1–12. https://doi.org/10.1016/J.COMPCHEMENG.2014.03.013

Anderson, J., 2009. How can the human mind occur in the physical universe?

Chen, H., Liu, S., Pang, L., Wanyan, X., Fang, Y., 2021. Developing an Improved ACT-R Model for Pilot Situation Awareness Measurement. IEEE Access 9, 122113–122124. https://doi.org/10.1109/ACCESS.2021.3108438

Deng, C., Cao, S., Wu, C., Lyu, N., 2019. Modeling Driver Take-Over Reaction Time and Emergency Response Time using an Integrated Cognitive Architecture: https://doi.org/10.1177/0361198119842114 2673, 380–390. https://doi.org/10.1177/0361198119842114

Huey, B., Wickens, C., 1993. Workload transition: Implications for individual and team performance.

Just, M.A., Carpenter, P.A., 1976. Eye fixations and cognitive processes. Cognitive Psychology 8, 441–480. https://doi.org/10.1016/0010-0285(76)90015-3

Kluge, A., Nazir, S., Manca, D., 2015. Advanced Applications in Process Control and Training Needs of Field and Control Room Operators. https://doi.org/10.1080/21577323.2014.920437 2, 121–136. https://doi.org/10.1080/21577323.2014.920437

Manca, D., Nazir, S., Colombo, S., Kluge, A., 2014. Procedure for automated assessment of industrial operators. Chemical Engineering Transactions 36, 391–396. https://doi.org/10.3303/CET1436066

39e Investigating Model Predictive Control for a Wafer Etch Temperature Control System Using Computational Fluid Dynamics Simulations

Henrique Oyama, Kip Nieman, Helen Durand

Model predictive control [1] has been extensively utilized in the process industries for decades, and in the academic literature it has typically been investigated for industrially-relevant processes using ordinary differential equations or their discrete-time equivalents. A relatively new trend in the process systems engineering literature has been exploring control of process systems using computational fluid dynamics (CFD) or finite element analysis (FEA) simulations. For example, [2] performed closed-loop simulations of a steam methane reforming reactor under proportional-integral control and under an optimization-based strategy. [3] expanded this work and used a model predictive controller considering a linear dynamic model. Recent work [4] has modulated the temperature at the wall of a 1-m version of a steam methane reforming reactor and simulated not only the impacts of the modulated input on process variables, but also on the maximum equivalent stress in the reforming tube walls. [5] has shown the impacts of run-to-run control on the performance of a plasma-enhanced chemical vapor deposition process. These works have shown the value of using CFD simulation to evaluate the performance of complex process systems under various control concepts or operating policies.

Given the successes in using CFD for control evaluation above, we evaluate another complex process with many interacting variables using CFD simulation. In this case, we focus on a temperature control system for a silicon wafer etching process, where heat is removed from two tanks and flows leaving these tanks are either diverted back to the tanks or mixed together to maintain the temperature at the inlet to a wafer etching process electrostatic chuck (ESC) at a required specification. The fluid leaving the ESC is then returned to the tanks. The manipulated inputs are valve positions for valves that return fluid to the tanks; this causes the temperatures observed throughout the system to be coupled so that designing a model predictive control algorithm for this system would be expected to be beneficial for dealing with the interactions between variables, as well as constraints such as the cooling tank heat removal rate limits. Though ordinary differential equations could be developed to describe this system, they do not permit the impacts of different control actions on the details of heat rate and flow profiles in the system to be examined. Therefore, we develop a CFD model of this full system, using several simplifications such as not explicitly modeling the valves or wafer etching process and instead incorporating their effects on the system transport profiles via user defined functions that set flow rate boundary conditions out of each diverter valve and into the next set of piping, or using an assumed rate of heat added to the cooling fluid by the outlet of the ESC.

We use the CFD simulations to develop first-order-plus-dead-time models for capturing input-output relationships throughout the system (e.g., relationships between estimated average temperatures or flow rates and the measured values, and relationships between valve positions and flow rates) and then use these to develop an ordinary differential equation model for tuning baseline proportional-integral-derivative (PID) controllers for the full system, which are then simulated in the CFD environment. Mesh and time independence tests are used to validate the transient simulation behavior and to simulate the system under the baseline control design consisting of several PID type controllers used for manipulating rates of heat removal at the cooling tanks and valve positions throughout the system. Subsequently, the first-order-plus-dead-time models are used in the design of a model predictive controller for the system. This controller is then implemented on the full CFD temperature control system, and the details of the transport profiles different under the different control strategies are compared.

References:

[1] Qin, S. J., andamp; Badgwell, T. A. (2003). A survey of industrial model predictive control technology. Control Engineering Practice, 11, 733-764.
[2] Lao, L., Aguirre, A., Tran, A., Wu, Z., Durand, H., andamp; Christofides, P. D. (2016). CFD modeling and control of a steam methane reforming reactor. Chemical Engineering Science, 148, 78-92.
[3] Wu, Z., Aguirre, A., Tran, A., Durand, H., Ni, D., andamp; Christofides, P. D. (2017). Model predictive control of a steam methane reforming reactor described by a computational fluid dynamics model. Industrial andamp; Engineering Chemistry Research, 56, 6002-6011.
[4] Nieman, K., D. Messina, M. Wegener and H. Durand, "Cybersecurity and Dynamic operation in Practice: Equipment Impacts and Safety Guarantees," Journal of Loss Prevention in the Process Industries, submitted.
[5] Crose, M., Zhang, W., Tran, A., Orkoulas, G., andamp; Christofides, P. D. (2018). Run-to-Run Control of Film Thickness in PECVD: Application to a Multiscale CFD Model of Amorphous Silicon Deposition. In Computer Aided Chemical Engineering (Vol. 44, pp. 511-516). Elsevier.

156a Keynote Talk-Bayesian Optimization for Additive Manufacturing of Thermoelectric Materials and Devices

Alexander Dowling, Ke Wang, Mortaza Saeidijavash, Minxiang Zeng, Tengfei Luo, Yanliang Zhang

Discovering and manufacturing functional materials with precisely tuned properties is a central goal of material science and engineering; yet materials discovery and additive manufacturing optimization is often slow and expensive. For example, novel solid-state thermoelectric (TE) materials have the potential to improve energy efficiency by converting waste heat into electricity. However, the performance of many state-of-the-art TE materials remains inadequate for adoption beyond niche applications. Current efforts to optimize photonic sintering, an important step in additive manufacturing of TE devices, rely on expert-driven trial-and-error search which is often extremely time-consuming and without the guarantee of improvement.

In this talk, we argue Bayesian optimization (BO) and related machine learning frameworks offer principled approaches to intelligentially recommend optimized experimental conditions by balancing exploitation and exploration. Specifically, we show successfully application of BO to materials composition (ink formulation), printing, and sintering steps in the additive manufacturing of n-type and p-type TE materials. We highlight the role of 'systems thinking' to synergistically fuse human intuition and BO machine learning models, ultimately resulting in the synthesis of record-breaking TE materials with under fifty experiments.

91a Probabilistic Machine Learning Based Soft-Sensors for Product Quality Prediction in Batch Processes

Max Mowbray, Philip Martin, Dongda Zhang

Statistical machine learning algorithms have been used to analyse industrial data for batch process monitoring and control. The use of statistical learning algorithms has been preferred because of the difficulties in identifying first principles predictive models for these processes. For example, in batch production of formulated products, the goal is typically to keep the batch on specification. However, measurement of key product qualities is often conducted offline, due to the expense of advanced process analytical technology. This generally means one measurement of key product quality is produced from a given batch. This provides difficulty for modelling end-product qualities, as one seeks to find a function mapping from temporal measurements of batch process variables (such as pressures, temperatures and flowrates) to one measurement of end-quality. This data is highly dimensional, multicollinear and nonstationary, which provides challenge to all model construction frameworks.

Recent research has considered the development of soft sensors, based on latent variable models, which automatically quantify the uncertainty associated with their prediction. In this study, we develop a three-step methodology to identify, visualize and systematically reduce data dimensionality for the construction of robust soft-sensors for end-product quality prediction. The approach first screens the entire dataset to identify critical time regions and operational variables, which correlate strongly with the end product quality one wishes to model. We then adopt multiway latent variable modelling to construct a latent space descriptive of the existing batches. Nonlinear estimators are then constructed from the reduced latent space to estimate final product quality, which are able to express model uncertainty automatically. Specifically, in this study, we explore the performance of Gaussian processes (GP), Bayesian neural networks (BNN) and heteroscedastic noise neural networks (HNN).

To highlight the efficiency and practical benefits of the approach, an industrial consumer goods product manufacturing process is presented as an example. The soft sensors were constructed and their performance was assessed via cross validation, using metrics which not only quantify the mean error in prediction, but also the quality of the uncertainty prediction. This assessment was informed with knowledge of the standard error associated with product end-quality measurement. The accuracy, reliability, and interpretability of the soft-sensors is discussed, tested and shown to generalise well on two test datasets. Specifically, the GP and HNN both provide good performing models, with accurate predictions in the mean (10% mean absolute percentage error, MAPE, in cross validation) and uncertainty estimates which reflect the standard error of measurement. Whereas the BNN, has low MAPE, but provides overly confident predictions, relative to the standard error of measurement. This result is also observed in assessment of the soft-sensors on the test set.

Innovations of this approach include ease in data visualisation, ability to identify major operational activities within the factory offline, as well as robustly and interpretably predict product end-quality.

156c Application of Artificial Intelligence to Predict Surfactant Adsorption

Shams Kalam, Sidqi Abu-Khamsin, Muhammad Shahzad Kamal, Shirish Patil, Syed Hussain, Emad W. Al Shalabi

Surfactants play a vital role in chemical enhanced oil recovery (cEOR). The quantity of surfactant loss due to adsorption on a rock directly influences a cEOR project's economics. Therefore, surfactant adsorption quantification is an important area of interest. Surfactant adsorption is influenced by several parameters, which makes its estimation difficult through analytical modeling. In this paper, a real experimental data set was used to predict surfactant adsorption as a function of minerals' percentage, surfactant equilibrium concentration, and maximum adsorption capacity. Several pure minerals were used to find the static adsorption of a Gemini surfactant using high-performance liquid chromatography (HPLC).

Artificial neural networks (ANN), eXtreme Gradient Boosting (XGBoost), and Random Forest (RF) were applied to build a digital model to estimate surfactant adsorption based on real static adsorption data. The data was divided into an 80:20 ratio for training and testing, respectively. The coefficient of determination (R2) and root mean squared error (RMSE) were used to find the optimal results after sensitivity analysis of hyperparameters. ANN outperformed XGBoost and RF. Feed-forward neural networks consisting of a single hidden layer with four hidden layer neurons gave the best results. The trained ANN model showed good agreement with the unseen data. This research reports a new digital model to predict surfactant adsorption using ANN as a function of surfactant concentration, maximum adsorption capacity, and mineral composition. The newly developed ANN model will save a lot of time in running tedious experiments and will provide a good quick estimate of surfactant adsorption.

156d A Deep-Learning Model of a Crude Distillation Unit

Jiannan Zhu, Feng Qian, Chen Fan, Vladimir Mahalec

Large amounts of data collected daily in refining plants are a rich source of information about the performance of the processing units. Using that data to create models that can be used to monitor and optimize operation can lead to significant benefits from improvements of the plant operation. In the last decade, numerous new deep learning techniques have emerged, especially in the fields of computer vision and natural language processing[1-4]. Such techniques have a potential to deliver improved data-driven models of refining and petrochemical process units. Recently, several data-driven models of the primary and secondary processing units in the refinery have applied the deep-learning methods[5-7]. However, most of these works have used the typical feed-forward neural network (FNN) to make some simple predictions.

This work introduces a new deep neural network structure for modeling the primary processing unit in the refinery, including the pre-flash tower, atmospheric tower, and vacuum tower. The model predicts properties of the products. Inspired by the work presented by Song in 2020[8], the model structure in this work contains two parts. The former part is the "figure generation" part, which converts the input variables into a two-dimensional "figure." The latter part is the convvolutional neural network (CNN), which processes these "figures" and generates predictions. Since the size of the input for the CNN is preferably a square, the number of the input variables will not form a proper square exactly. There are two ways to generate the suitable input "figures" for CNN. One is to increase the dimensionality that maps the input into a larger square-sized "figure." Another way is to map the input into a smaller square-sized "figure" through some dimensionality reduction methods. Both ways have been evaluated by using SOM (self-organizing-map) and PCA (Principal Component Analysis), respectively. The results showed that the model using PCA performs better. In the CNN part of the model, we have used the ResNet structure proposed by He et al. in 2015[1]. It can be seen in Figure 1 that the CNN (ResNet) part consists of several residual blocks. Each block contains a convolutional layer, batch normalization layer, activation function, and an additional convolutional layer to identity mapping the input data. The residual blocks, which add the link between the input and the output, can enhance the convergence and accuracy as well as ensure that the structure will not be overfitted easily.

In this work, several different sets of features (input variables) have been investigated. Firstly, the effect of different selection schemes of input variables on the accuracy of prediction results was investigated. It has been found that, compared with the traditional way of selecting the input variables, adding some ratios based on prior knowledge, e.g. as) and others, as well as selected tray temperature leads to a much more accurate model. In addition, different sizes of the input "figures" for the ResNet part have been tested. Comparison experiments have been carried out based on the backpropagation (BP), PCA-BP, SOM-ResNet, and PCA-ResNet models. In addition, the minimum amount of data needed to support the different models have also also investigated.

Among the several models, the PCA-ResNet model has the best prediction accuracy and a smoother and more convergent error decline curve.

References

1. He, K., et al., *Deep Residual Learning for Image Recognition*, in *2016 IEEE Conference on Computer Vision and Pattern Recognition (CVPR)*. 2016. p. 770-778.
2. Ioffe, S. and C. Szegedy. *Batch normalization: Accelerating deep network training by reducing internal covariate shift*. in *International conference on machine learning*. 2015. PMLR.
3. Krizhevsky, A., I. Sutskever, and G.E. Hinton, *ImageNet Classification with Deep Convolutional Neural Networks*. Communications of the Acm, 2017. **60**(6): p. 84-90.
4. Simonyan, K. and A. Zisserman, *Very Deep Convolutional Networks for Large-Scale Image Recognition*. Computer Science, 2014.

5. Ochoa-Estopier, L.M., M. Jobson, and R. Smith, *Operational optimization of crude oil distillation systems using artificial neural networks.* Computers andamp; Chemical Engineering, 2013. **59**: p. 178-185.
6. Shin, Y., R. Smith, and S. Hwang, *Development of model predictive control system using an artificial neural network: A case study with a distillation column.* Journal of Cleaner Production, 2020. **277**: p. 124124.
7. Rosli, M. and N. Aziz. *Review of neural network modelling of cracking process.* in *IOP Conference Series: Materials Science and Engineering.* 2016. IOP Publishing.
8. Song, W., et al., *Modeling the Hydrocracking Process with Deep Neural Networks.* Industrial andamp; Engineering Chemistry Research, 2020. **59**(7): p. 3077-3090.

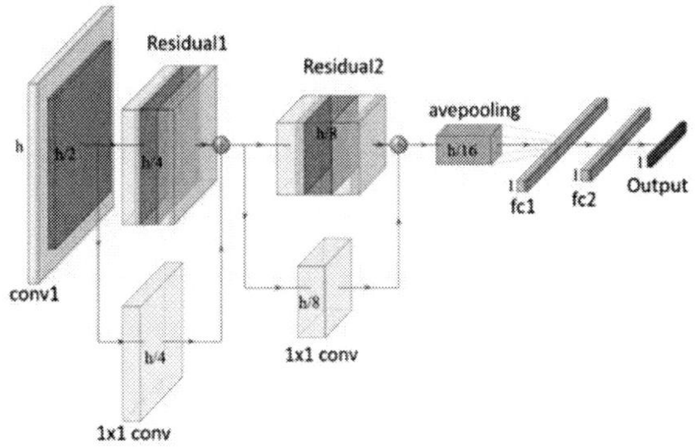

156e The Effect of Chemical Representation on Active Machine Learning Towards Closed-Loop Optimization

Alexander Pomberger

Multivariate chemical reaction optimization involving catalytic systems is a non-trivial task due to the high number of tuneable parameters and discrete choices. Closed-loop optimization featuring active Machine Learning (ML) represents a powerful strategy for automating reaction optimization. However, the translation of chemical reaction conditions into a machine-readable format comes with the challenge of finding highly informative features which accurately capture the factors for reaction success and allow the model to learn efficiently. Herein, we compare the efficacy of different calculated chemical descriptors for a high-throughput generated dataset to determine the impact on a supervised ML model when predicting reaction yield. Then, the effect of featurization and size of the initial dataset within a closed-loop reaction optimization was examined. Finally, the balance between descriptor complexity and dataset size was considered. Ultimately, tailored descriptors did not outperform simple generic representations, however, a larger initial dataset accelerated reaction optimization.

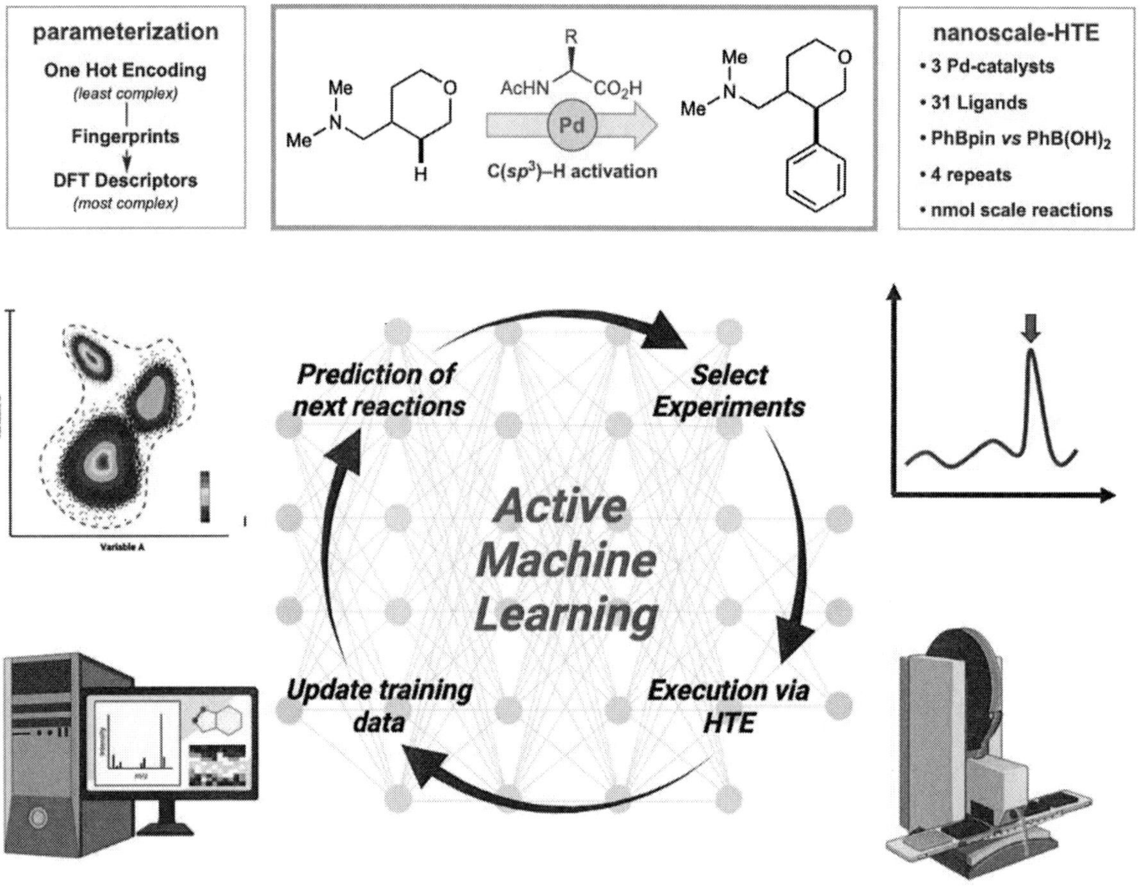

156f Neural Network-Based Automated Detection of Functional Groups in Spectroscopic Data- Bridging the Gap in Online Monitoring of Complex Reaction Systems

Karthik Srinivasan, Vinay Prasad

The problem of reliable monitoring of complex reacting systems is largely dependent on accurate enumeration of the reacting species and the reactions governing their conversion. However, the absence of information, as it is common in the case of distributed feedstocks such as bitumen and biomass, makes monitoring reliant on (i) molecular-level data from process analytical tools like spectroscopic sensors, and (ii) high physical fidelity statistical models that use the process data to infer the reaction mechanism structure. Understanding the actual chemistry in the reactions of such complex feedstocks without *a priori* knowledge of the system becomes intractable due to the expanse of chemical reaction space. In the case of system where both reactions and species are known, the development of the reaction network becomes a simple task of 'applying' the reactions to the species based on expert knowledge. In contrast, systems for which both reactions and species are unknown, require mathematical manipulations of the spectroscopic data to decipher the reaction network topology. Structure preserving non-negative tensorial factorization of the time evolving spectroscopic data results in latent factors that can be projected on to the spectral mode to be interpreted as the spectra of the reactive pseudo-components. Domain knowledge can then be employed to map the spectral peaks to signatures of active functional groups in each component [1].

We present an automated technique to infer information about functional groups in a spectrum without prior knowledge of the species. The identification problem is cast as a classification problem and is solved using an artificial neural network. The network architecture incorporates 1-dimensional convolution blocks to extract features corresponding to a specific functional group. The convolution block slides filters of predetermined size along the wavenumber axis. The convolution operation between the filter and the input spectra generates localized regions of 'hot spots' indicating a match between the two. The usage of a 1-D convolution block restricts this operation to just one axis, therefore making the classifier absorbance invariant. According to Beer's Law, the height of a peak in spectrum is dependent on the concentration, path length and the absorptivity of the compound in that particular region. In the case of complex reactive mixtures, the concentration profiles of the species vary over time thus engendering a need of absorbance invariance in an automated detection scheme. Previous works have established a similar routine for automated functional group detection using Convolutional Neural Networks (CNNs) but incorporate a 2-D convolution operation, thus making the system dependent on the peak height [2]. The convolution blocks are followed by a fully connected layer to analyse the features from the convolutional layers. The output of the network is a bit vector of size 13 indicating the presence or absence of a particular functional group in the species with a sigmoidal activation layer being used to calculate the probabilities of each functional group. The network is trained to simultaneously predict the presence of different functional groups and hence training the neural network becomes a task of learning the joint probability distribution for each functional group in the presence of others [3]. This task is commonly referred to as multi-label classification, where an input can belong to multiple classes simultaneously. In this regard, the training is performed using backpropagation with the binary cross entropy loss for each bit in the output. Hamming loss is used as a test metric to quantify the degree of accuracy of the classifier. The Hamming loss can be defined as the fraction of incorrect predictions over the label size. In order to improve the confidence on the predictions, a reconstruction metric is also considered. The reconstruction metric computes the difference between the input spectrum and a theoretical spectrum based on the predictions of the classifier. This allows for the analysis of the feature space, improving the fidelity of the model.

The network hypothesis generation is then carried by learning a graphical structure from the resolved pseudo-components using Bayesian structure learning algorithms. The graph indicates the transition of one species to another, thereby providing the skeletal structure of the network where the nodes of the graph correspond to each pseudo- component whose spectral signature have been identified by the classifier. The generation of the reaction graph, also provides a validation metric for the classification. This protocol has been applied to describe the reaction network for the hydrous pyrolysis of biomass and results indicate a high level of correspondence with the reactions described in the literature thus providing the ability to perform online monitoring with interpretability without any prior knowledge of the system.

Keywords: Convolutional neural networks, reaction hypothesis, latent factor projection, Bayesian networks, reaction monitoring

156g Effects of Noise When Implementing Linear Control Laws on Quantum Computers

Keshav Kasturi Rangan, Kip Nieman, Helen Durand

Quantum computation is receiving greater interest in process systems engineering. For example, it has been used for optimization [1] and fault detection [2]. From a controls perspective, however, quantum computing remains relatively unexplored within the process systems engineering field. One of the reasons for this is that it is not currently clear whether there exists a quantum computing algorithm that can provide speedups for control design to provide a benefit from quantum computing. However, it is our hypothesis that a more detailed understanding of how control design and theory interact with quantum computing algorithms could help to suggest modifications to the logic of the quantum computing algorithms to achieve control objectives, which may also aid in revealing beneficial algorithms from a computational perspective. This work thus takes a control-theoretic perspective on quantum computing for control, rather than an algorithmic efficiency perspective, with the hope of spurring future work in algorithm development.

In this talk, we will describe our work at the interface of control and quantum computing,[3] focusing first on the impacts of noise in quantum hardware on the stabilization of the state of a single-input/single-output process under a control law that can be computed using an addition circuit based on the Quantum Fourier Transform (QFT).[4] This system was simulated using a quantum simulator available from IBM's Quantum Experience. This allows addition of a custom noise model to simulate errors in the computations due to the hardware that would be seen on an actual device. The custom noise model used was based on the depolarizing error. The state and input trajectories for the process when the control law is the same but control actions are computed by a classical computer or the quantum simulator in the absence of noise are compared with the case that the inputs are computed by the quantum simulator with noise. This is performed for different initial conditions and numbers of runs of the quantum simulator. In all cases, the closed-loop state was eventually driven to the origin. This motivates further study of the manner in which different tuning variables for the system impact the result. For example, different error parameters were also tested, and both an estimate of the percentage of the inputs that would drive the closed-loop state toward the origin from a given state as well as the different frequencies with which these inputs were computed for different implementations of the control law at the addition algorithm implementation level were examined.

The fact that undesired inputs were still able to drive the closed-loop state toward the origin with time under the control policy with different error parameters in the noise model for the quantum device raises the question of what this result implies for cybersecurity of control systems. Our group has developed control-theoretic studies regarding cybersecurity of process control systems (e.g., [5,6]) in which we develop conditions under which we can guarantee that attacks are detected when the process is controlled by an advanced control law, and/or when we can guarantee safety until the time of detection in the sense that the closed-loop state is maintained in a bounded region of state-space. Cyberattacks on a control system can be thought of as an undesired input policy; we therefore compare the results under the quantum computing and cyberattack scenarios to better understand relationships between these cases and what the relationships indicate about techniques for handling both methods. Finally, we conclude with a discussion of techniques for examining quantum noise in the context of other control laws besides that applied for the single input-single output control law.

References:

[1] Ajagekar, A., andamp; You, F. (2019). Quantum computing for energy systems optimization: Challenges and opportunities. *Energy*, *179*, 76-89.

[2] Ajagekar, A., andamp; You, F. (2020). Quantum computing assisted deep learning for fault detection and diagnosis in industrial process systems. *Computers andamp; Chemical Engineering*, *143*, 107119.

[3] Nieman, K., Rangan, K. K., andamp; Durand, H. Control Implemented on Quantum Computers: Effects of Noise, Non-Determinism, and Entanglement. Industrial andamp; Engineering Chemistry Research (in press).

[4] Ruiz-Perez, L., andamp; Garcia-Escartin, J. C. (2017). Quantum arithmetic with the quantum Fourier transform. *Quantum Information Processing*, *16*(6), 1-14.

[5] Rangan, K. K., Oyama, H., andamp; Durand, H. Integrated Cyberattack Detection and Handling for Nonlinear Systems with Evolving Process Dynamics under Lyapunov-based Economic Model Predictive Control. Chemical Engineering Research and Design (2021).

[6] Oyama, H., Rangan, K. K., andamp; Durand, H. Lyapunov-Based Economic Model Predictive Control for Detecting and Handling Actuator and Simultaneous Sensor/Actuator Cyberattacks on Process Control Systems. Frontiers in Chemical Engineering, 11.

575a Multiscale Healthcare Supply Chains: Patient in the Loop

Miriam Sarkis, Niki Triantafyllou, Andrea Bernardi, Nilay Shah, Maria Papathanasiou

Personalised therapies offer ground-breaking opportunities in the treatment of life-threatening diseases, including cancer. Currently, clinically successful personalised therapies involve infusing previously engineered cells to the target patient; those are then used to produce therapeutic proteins and/or other factors which serve as the treatment itself. In the instance of autologous therapies, the patient's own tissue/cells are collected, modified and re-administered to the same patient [1][2]. The patient-specific nature of autologous therapies makes them a unique class of therapeutics to manufacture and distribute, coupled with a consumer-centric pharmaceutical supply chain [3].

Manufacturers in the space of personalised medicine must carefully orchestrate resources to ensure product availability in the right place and at the right time. This translates into: (a) patient-specific batches and distribution nodes and (b) challenges in demand forecasting which impact raw material planning activities. This is further highlighted for next-generation vaccines and therapeutics, which use viral vectors as means to genetically modify the therapy in-manufacturing. Vectors are equally coupled with a lengthy and complex production process, leading to a interconnected network of manufacturing and distribution of complex biologics.

Furthermore, individual product orders vary on a day-to-day basis and short product shelf-life restricts waiting and storage times. This highlights the need for responsive and scalable supply chains that readily adapt to demand dynamics. Constraints on the flexibility of the manufacturing process are also imposed by strategic decisions, which range from mid-term acquisition of raw materials to long-term investments in additional capacity and infrastructure. In this space, computer-aided tools can support supply chain design and planning under uncertainty, whilst capturing interconnectivity of strategic and operational decision-making processes [4].

We propose a multiscale mixed-integer linear programming (MILP) framework for the design and optimisation of end-to-end CAR T cell therapy supply chains (Figure 1). Given a set of location-specific product demands, candidate location for network nodes (e.g. hospitals, manufacturing facilities and Quality Control sites), capital and operating expenditures, the optimisation determines suitable candidate supply chain structures, investment plans, production schedules and distribution plans. Production and inventory levels of required raw materials are also suggested by the optimisation algorithm. The candidate networks are assessed for cost, responsiveness and resilience with respect to the uncertain demand, as well as unexpected delays in the end-to-end product lifecycle.

[1] Novartis (2018), KYMRIAH Treatment Process, Dosing andamp;amp; Administration | HCP.

[2] Kite Pharma (2018), First CAR T Therapy for Certain Types of Relapsed or Refractory B-Cell Lymphoma.

[3] Papathanasiou, M.M.; Stamatis, C.; Lakelin, M.; Farid, S.; Titchener-Hooker, N.; Shah, N. (2020). Autologous CAR T-cell therapies supply chain: challenges and opportunities? *Cancer Gene Therapy*, 27, 799-809.

[4] Sarkis, M.; Bernardi, A.; Shah, N.; Papathanasiou, M. (2021). Decision support tools for next-generation vaccines and advanced therapy medicinal products: present and future, *Current Opinion in Chemical Engineering*, 32, 100689.

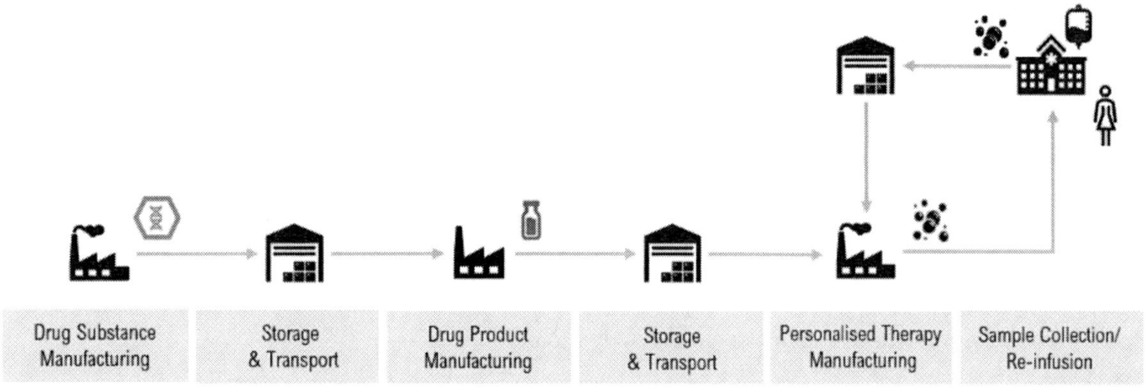

Figure 1: General schematic of end-to-end personalised healthcare supply chains.

575b Knowledge-Informed Data-Driven Modeling for Robust Prediction of Microbial Inactivation in Food

Firnaaz Ahamed, Steve Zhang, Hyun-Seob Song

Prevention of the growth of harmful pathogens or spoilers in food products is an important requirement for ensuring food safety and quality. Predictive mathematical models can serve as a useful tool for optimizing microbial inactivation processes in food. A common interest is to identify a functional dependency of the *D*-value (i.e., the time for a microbial population to shrink to 10% of its initial level) on a variety of potential factors, including temperature, pH, water activity, etc. While polynomial functions are typically used for this purpose, the empirical choice of the terms often results in underfitted or overfitted models that lack accuracy and robustness in predictions. This problem occurs when the model includes fewer terms than required or more terms than needed, requiring a systematic approach to determine optimal model structures by accounting for a trade-off model accuracy vs. complexity. In this work, we apply an advance data-driven approach - termed Sparse Identification of Nonlinear Dynamics (SINDy) - to address this issue. SINDy enables automatic discovery of model equations through a parsimonious selection of appropriate terms from a pre-built library. Through case studies using various literature data, we demonstrate how to systematically develop microbial inactivation models using the data-driven modeling approach and how to improve model accuracy by choosing appropriate functional forms of basic input and output variables by leveraging experts' knowledge (i.e., based on the analysis of associated mechanistic models here). Beyond enhanced quantitative data fit, we generalize the workflow to systematically identify models without assuming *a priori* model structure and robust stepwise model tuning through a balanced compromise between accuracy and complexity as gauged by the Akaike information criterion. We also integrate global sensitivity analysis with the optimized model to evaluate the impacts of individual factors (and their interactions thereof) on microbial inactivation. Beyond significantly facilitating the model building process for optimizing, the proposed workflow enables systematic identification of individual and combined impacts of key variables/terms on the target variable (i.e., *D*-value).

575c Systematic Design of Solvent Recovery Systems in Pharmaceutical Processes

Jake Stengel, Austin Lehr, Emmanuel Aboagye, John Chea, Kirti Yenkie, David Streather, Michael Parker, Claire MacLeod, Peter Schell

The global pharmaceutical market is projected to nearly double in size between the years of 2020 and 2028[1]. This growth has raised concerns around the sustainability of the industry as a majority of pharmaceutical products are formed through organic synthesis routes that require many sequential reaction steps, and large quantities of multiple organic solvents. Product formation processes are followed by downstream operations such as numerous solid-liquid separation and purification steps [2], [3] that generate waste streams containing organic solvents since they are not a part of the reaction stoichiometry. It is common in the pharmaceutical industry for solvents to account for 80-90% of the mass intensity for the process and due to the stringent purity requirements for solvents, these are often discarded after a single use.[4]–[6]. A common method for estimating the "greenness" of a process is to calculate the E-factor which is a measure of the mass of waste generated per mass of product. Due to the various sources of waste generation, the API (Active Pharmaceutical Ingredient) manufacturing process has an E-factor (mass of waste per mass of product) usually between 25 and 100 [7]. Current leaders in the Pharmaceutical industry like AstraZeneca and Glaxo-Smith-Kline (GSK) have implemented Life Cycle Assessment (LCA) metrics to evaluate the emissions associated with their process. It has become important to identify and optimize the "hot spots" of these processes in relation to their life cycle. In the case of the Pharmaceutical industry one such hot spot is the solvent consumption, with the current research being into selecting "greener" solvents. The focus has been put on obtaining higher reaction yields rather than developing systems that can recover, purify, and reuse solvents and thus reduce the amount of fresh solvent required in a process.

This work proposes the development of roadmap and associated software tool for the systematic design of solvent recovery processes that emphasize the importance of physical properties and stream composition-based resource conservation and recovery. The solvent recovery roadmap is generated using a stage-wise superstructure approach consisting of technology models. Each technology is modeled as a set of mass/energy balance and design equations. These models are then formulated as a mixed integer non-linear programming (MINLP) problem in General Algebraic Modeling System (GAMS). By implementing binary variables, used to make "yes" or "no" decisions, we are able to select a path that is simultaneously optimized for minimum cost and environmental impact while meeting quality requirements. The associated software tool is created using MATLAB to generate a graphical user interface (GUI). This GUI is used to allow the user to define the components, composition, and desired quality specifications and display the relevant reports from the GAMS model. These reports include key performance indicators (KPIs) in the pharmaceutical industry such as carbon footprint, process mass intensity (PMI), water usage, and cost per kilogram of solvent recovered. To test the validity of this application, a series of case studies from a major pharmaceutical company were analyzed. The complexity of these case studies ranged from a simple binary system to a five-component system with multiple azeotropic mixtures.

[1] "Pharmaceutical Manufacturing Market Size Report, 2021-2028." https://www.grandviewresearch.com/industry-analysis/pharmaceutical-manufacturing-market (accessed Mar. 07, 2022).

[2] R. G. Harrison, "Bioseparation Basics," *Chemical Engineering Progress*, Oct. 2014. Accessed: Jun. 14, 2016. [Online]. Available:

http://www.aiche.org/resources/publications/cep/2014/october/bioseparation-basics

[3] P. A. Belter, E. L. Cussler, and W.-S. Hu, *Bioseparations: downstream processing for biotechnology*. USA: Wiley, 1988.

[4] C. Jimenez-Gonzalez, C. S. Ponder, Q. B. Broxterman, and J. B. Manley, "Using the Right Green Yardstick: Why Process Mass Intensity Is Used in the Pharmaceutical Industry To Drive More Sustainable Processes," *Org. Process Res. Dev.*, vol. 15, no. 4, pp. 912–917, Jul. 2011, doi: 10.1021/op200097d.

[5] M. J. Raymond, C. Stewart Slater, and M. J. Savelski, "LCA approach to the analysis of solvent waste issues in the pharmaceutical industry," *Green Chem.*, vol. 12, no. 10, pp. 1826–1834, 2010, doi: 10.1039/C003666H.

[6] R. K. Henderson *et al.*, "Expanding GSK's solvent selection guide – embedding sustainability into solvent selection starting at medicinal chemistry," *Green Chem.*, vol. 13, no. 4, pp. 854–862, 2011, doi: 10.1039/C0GC00918K.

[7] R. A. Sheldon, "The E Factor: fifteen years on," *Green Chem.*, vol. 9, no. 12, pp. 1273–1283, 2007, doi: 10.1039/B713736M.

575d Quantifying the Intra-Sample Variability of Gelatinization of Starch Granules for Optimization of Food Processing

Lanxin Mo, James Cheon, John Frostad

Starch is a critical ingredient in determining the taste and texture of many food products. In an optimal process the starch should be fully gelatinized, because undercooked starch has a characteristic chalky flavor that is undesirable. On the other hand, overcooking can decrease the nutrition and unnecessarily increase the processing time. In this work, we present results from a "Particle Cohort Study" (ParCS) to quantify the gelatinization process of large number of individual starch granules from four common legume starches. For the first time, we quantify the intra-sample variability of individual granules, as this information is not available from the standard techniques: starch pasting and differential scanning calorimetry. The results indicate that these legume starches have a remarkably low variability in gelatinization properties, which could be advantageous for food processing. We also demonstrate that a simple empirical model is highly effective at predicting the rate and extent of gelatinization. These results are expected to enable increased optimization of food processing involving starch granules.

575e Hybrid Modeling Using Universal Differential Equations for Lab-Scale Batch Production of β-Carotene Using *Saccharomyces Cerevisiae*

Mohammed Saad Faizan Bangi, Katy Kao, Joseph Kwon

Carotenoids are a diverse group of yellow-orange pigments present in many biological systems, and are produced by bacteria, fungi, and plants. They have been extensively used as constituents in dietary and vitamin supplements, and food pigmentation because of their colored characteristics. Specifically, β-carotene plays an important role as a precursor to Vitamin A. Additionally, it has been shown that β-carotene has a positive impact on human health with its antioxidant and protective properties against cancer. Currently, the use of chemical technology for the production of carotenoids leads to byproducts with undesirable effects upon consumption due to which the use of microbial sources for their production has received lots of attention [1]. But modeling the production of β-carotene in a fermentation process using first-principles such as conservation of mass and energy, kinetic laws, thermodynamic laws, and transport laws, etc., leads to limited accuracy. This is because first-principles cannot fully account for all the complex phenomena that occur within the process unless multiple laws are used which can lead to high computational costs [2]. Alternatively, data-driven modeling can be used but it suffers from poor extrapolation properties. To overcome these limitations, hybrid modeling is utilized which combines the first-principles and data-driven modeling, specifically deep neural networks, approaches resulting in superior accuracy and extrapolation properties than first-principles models and data-driven models, respectively [3].

Recently, a new class of neural networks called Neural ODEs has been developed wherein instead of specifying a discrete sequence of hidden layers in the neural network structure, the progression of the input through the hidden layers becomes continuous and is represented using an ODE [4]. Solving this ODE using a black-box ODE solver gives the output of the Neural ODEs. These continuous-depth Neural ODE networks have a constant memory cost, are able to adapt their evaluation strategy to each input, and explicitly trade computational speed for accuracy. These Neural ODEs can be combined with a first-principles-based model to build a hybrid model called Universal Differential Equations (UDEs) [5]. In this work, we built a UDE-based hybrid model for batch production of β-carotene using *Saccharomyces cerevisiae* strain mutant SM14. This model was developed using multiple software packages in Julia programming language and trained using data obtained from experiments wherein the initial glucose concentration was 20 g/L. Subsequently, the trained UDE model was tested using another experimental dataset wherein the initial glucose concentration was 22.36 g/L. The UDE-based hybrid model shows accuracy superior to the existing first-principles model specifically in the case of biomass, acetic acid and β-carotene concentrations. This work illustrates that UDE-based hybrid models can be utilized to quantify the unknowns in the first-principles model thereby improving its overall accuracy.

Literature cited:

[1] Ordonez, M.C., Raftery, J.P., Jaladi, T., Chen, X., Kao, K., Karim, M.N. Modeling of batch kinetics of aerobic carotenoid production using *Saccharomyces cerevisiae*. *Biochem. Eng. J.*, **114**:226-236, 2016.

[2] Bangi, M.S.F., Kwon, J.S.I. Deep hybrid modeling of chemical process: application to hydraulic fracturing. *Comput. Chem. Eng.*, **134**:106696, 2020.

[3] Shah, P., Sheriff, M.Z., Bangi, M.S.F., Kravaris, C., Kwon, J.S.I., Botre, C., Hirota, J. Deep neural

network-based hybrid modeling and experimental validation for an industry-scale fermentation process: Identification of time-varying dependencies among parameters. *Chemical Engineering Journal,* 135643, 2022.

[4] Chen, T.Q., Rubanova, Y., Bettencourt, J., Duvenaud, D.K. Neural ordinary differential equations. *Advances in Neural Information Processing Systems 31*, 6571-6583, 2018.

[5] Rackauckas, C., Ma, Y., Martensen, J., Warner, C., Zubov, K., Supekar, R., Skinner, D., Ramadhan, A. Universal Differential Equations for Scientific Machine Learning. *arXiv:2001.04385*, 2020.

575f Machine Learning-Based Modeling and Predictive Control of Crystallization Processes Under Batch-to-Batch Parametric Drift

Yingzhe Zheng, Tianyi Zhao, Xiaonan Wang, Zhe Wu

Batch crystallization is typically regarded as one of the crucial unit operations in pharmaceutical manufacturing as more than 90% of the active pharmaceutical ingredients are synthesized in the form of crystals [1]. The selection of the optimal operating and control strategies is of great interest to pharmaceutical manufacturers in achieving a more effective and eco-friendlier attainment of the crystalline product specification targets (e.g., crystal size distribution (CSD), purity, shape), which in turn dictate the performance of downstream operations, and are hence the pivotal enablers for enhanced productivity, smoother downstream processing, and superior drug bioavailability. While model predictive control (MPC) has emerged as the state-of-the-art optimal control strategy for chemical processes, several limitations (e.g., high computational costs for first-principles model-based MPC, low modeling accuracy and deteriorating control performance with respect to batch-to-batch (B2B) parametric drift [2, 3]) have to be addressed for its practical implementation.

Motivated by the above considerations, this work proposes a general framework for constructing computationally efficient machine learning models as substitutions for the mechanistic models in MPC and developing machine learning (ML)-based MPC schemes equipped with error-triggered online model update mechanisms for implementation to batch crystallization processes under the influence of B2B parameter drift. Specifically, we consider a seeded fesoterodine fumarate cooling crystallization and dissolution in a batch crystallizer ([6]) and present the methodology for designing a ML-based MPC controller to optimize product yield, crystal size, and energy consumption while accounting for the physical constraints on cooling jacket temperature. The ML model is constructed via the integration of recurrent neural network (RNN) with autoencoder (AE), where the RNN is developed to model the process dynamics and the AE is incorporated as a dimensionality reduction technique to further enhance the computational efficiency [4, 5]. Deviations in the crystallization kinetic parameters are considered in the closed-loop simulations to account for B2B parametric drift, and error-triggered online update mechanisms are incorporated into the MPC to improve its control performance by taking the availability of real-time system measurements into account. The results demonstrate that the proposed AERNN-MPC with online update, irrespective of the accessibility to real-time measurement data, achieves a desired closed-loop performance as compared with the MPC without online update.

[1] A. J. Alvarez and A. S. Myerson. Continuous plug flow crystallization of pharmaceutical compounds. *Crystal Growth andamp; Design*, 10(5):2219–2228, 2010.

[2] Q. Su, M. Chiu, and R. Braatz. Integrated b2b-nmpc control strategy for batch/semibatch crystallization processes. *AIChE Journal*, 63(11):5007–5018, 2017.

[3] J. S. I. Kwon, M. Nayhouse, G. Orkoulas, D. Ni, and P. D. Christofides. A method for handling batch-to-batch parametric drift using moving horizon estimation: application to run-to-run MPC of batch crystallization. Chemical Engineering Science, 127, 210-219, 2015

[4] T. Zhao, Y. Zheng, J. Gong, and Z. Wu. Machine learning-based reduced-order modeling and predictive control of nonlinear processes. *Chemical Engineering Research and Design*, 179:435–451, 2022.

[5] Y. Zheng, X. Wang, and Z. Wu. Machine learning modeling and predictive control of batch crystallization process. *Industrial andamp; Engineering Chemistry Research*, in press, 2022.

[6] M. Trampuž, D. Teslić, and B. Likozar. Crystallization of fesoterodine fumarate active pharmaceutical ingredient: Modelling of thermodynamic equilibrium, nucleation, growth, agglomeration and dissolution kinetics and temperature cycling. *Chemical Engineering Science*, 201:97–111, 2019.

575g Models for Mammalian Cell Cultures Based on Long-Short Term Memory Recurrent Neural Networks

Satish Parulekar, Essa Almutar

Mammalian cell cultures have become the favored production hosts for MAbs and therapeutic proteins, since microbial systems are not able to carry out the complex post-translational and functional modifications of these proteins, such as glycosylation. Products of these cell cultures have broad applications in vaccination, drug screening and development, and gene therapy. Monoclonal antibodies (MAbs) are significant reagents used extensively in diagnostic assays, therapeutic applications, and in vivo imaging. Chinese Hamster Ovary (CHO) cells and hybridoma cells, which share similar metabolic characteristics, have been popular cell types for production of MAbs. The efficient performance of these cell cultures requires highly specialized culture media to enhance MAb yield for in vitro production in view of substantial cell death and reduced MAb productivity due to the variations in culture conditions. Although production practices have been employed for decades, cell kinetics is still under investigation to obtain quantitatively as well as qualitatively cost-effective production strategies. Creating these strategies requires understanding of cell metabolism affected by process dynamics in culture environments. Kinetic models empower us to illustrate quantitative cell growth and metabolic activity, which allows prediction of different cell phenotypes and provides better understanding of cell physiology, which is important in optimization of MAb production in animal cell cultures. Kinetic descriptions of mammalian cell cultures are difficult due to these complexities. Development of first principles models (FPMs) is tedious. While FPMs provide qualitative understanding of the cell culture processes, their applications in dynamic operations for prediction, optimization and feedback control of mammalian cell cultures is limited due to their rigid structure. A recent focus has been on using statistical methods for modeling mammalian cell cultures. This approach has limited success due to it being restrictive to particular strains. In this work, we introduce the Recurrent Neural Network Long-Short Term Memory (LSTM) method for modeling mammalian cell cultures. The LSTM method uses neural network algorithms to model and analyze sequential data and is capable of handling large and nonlinear experimental databases. This method has been improved in the past decade, and with the availability of large databases and better computing power, we are able to model and analyze mammalian cell cultures. The models developed here used data generated from in silico experiments involving a large number of process inputs and outputs. The model simulations predict the trends observed in batch and fed-batch mammalian cell cultures for key nutrients glucose and glutamine, viable cell density, target product (monoclonal antibody) titer, and inhibitory metabolites lactate and ammonia with high accuracy. The predictions can be improved by further tuning of the modeling platform. The results of this study illustrate the promise of the LSTM method in other bioprocess applications, including optimization of nutrient media for cell cultivation and synthesis of high value target metabolites. The LSTM method, together with reinforcement learning (RL), can be a backbone in control of bioprocesses. Since RL relies on using LSTM in its implementation, RL has immense potential in optimal process control of stochastic dynamic systems.

350a Control-Theoretic Considerations for Manufacturing in Space

Kip Nieman, Katie Tyrrell, Helen Durand

As commercial space travel becomes more common, it becomes increasingly important to consider how people might be able to survive in distant locations, such as on colonies on Mars or during deep space travel. Since supplies from Earth would take months to reach people in space, help from Earth could not be relied upon when rapid solutions to problems are required, such as during an emergency, which represents a huge reliability and logistics challenge. To handle such issues, it would be expected that the future of commercial space will usher in new manufacturing considerations for production in space [1, 2]. However, the first question to be addressed is what does the future of manufacturing in space look like. Delivering computing devices into space will likely increase expenses and decrease sustainability, as it adds cargo costs into space and back from space or, if computing devices are not returned to earth, it removes otherwise recyclable electrical components from earth's manufacturing system. Furthermore, if computing devices rapidly develop or are sensitive (e.g., quantum computers), consistently trying to replace computing devices in space used for manufacturing may not provide as streamlined or adaptive of a strategy as might be desired. We could ask if instead Cloud resources on earth might be tapped into to aid in manufacturing. However, the time delays involved present an issue. There have been prior studies on handling asynchronous or delayed measurements even in advanced control laws such as economic model predictive control [3]; however, delays in space are different because they involve both delay in the receipt of sensor measurements and actuator outputs.

In this work, we use several control-theoretic approaches to analyze the conditions under which manufacturing in space could take advantage of computations on earth. In one case, we utilize predictions of the process state both as the measurements reach the earth and as the control actions return to space to compute control actions. In another, we send a set of control actions back to the space manufacturing system and clarify what range of state measurements should lead the controller to select one of the control actions. We consider conditions under which doing this constant communication is an improvement to storing some set of control action-state measurement pairs in the space manufacturing system. We also discuss data privacy issues by discussing concepts which exist for securing Cloud computation (e.g., homomorphic encryption [4, 5]) or where redundancy might be used to aid in attempting to confuse an attacker.

[1] Owens, A., and De Weck, O., 2016. "Systems Analysis of In-Space Manufacturing Applications for the International Space Station and the Evolvable Mars Campaign." *AIAA SPACE 2016*. 5394.

[2] Makaya, A., Pambaguian, L., Ghidini, T., Rohr, T., Lafont, U., and Meurisse, A., 2022. "Towards out of earth manufacturing: overview of the ESA materials and processes activities on manufacturing in space." *CEAS Space Journal* pp.1-7.

[3] Heidarinejad, M., Liu, J., Christofides, P.D., 2012. "Economic model predictive control of nonlinear process systems using Lyapunov techniques." *AIChE Journal* 58, pp. 855-870.

[4] Morris, L., 2013. "Analysis of partially and fully homomorphic encryption." *Rochester Institute of Technology* pp. 1-5.

[5] Caroline, F. and Galand, F., 2007. "A survey of homomorphic encryption for nonspecialists." *EURASIP Journal on Information Security 2007* pp. 1-10.

350b Some of the Variables, Some of the Times, with Some Things Known: Identification with Partial Information

Saurabh Malani, Thomas Bertalan, Tianqi Cui, Michael Betenbaugh, Jose Avalos, Ioannis G. Kevrekidis

Experimental data often consists of variables measured independently at different sampling rates, such that the Δt between successive measurements of a variable is non-uniform, and that at a specific time point only a subset of all variables are sampled. Previous methods to learn dynamical systems from such data use interpolation, imputation and/or sub-sampling to re-organize or modify the training data prior to learning [1, 2]. We present an approach using a multi-layer perceptron embedded in a Runge-Kutta numerical integrator template to learn the right-hand side of the underlying governing ODEs [3]. This neural-network model is iterated step-wise through time on a batch of trajectories with loss calculated at the time points data is available, allowing for learning from data sampled at arbitrary time points *without* data modification. Finally, we show how this network can be integrated with physics-informed 'grey-boxes' to learn kinetic rate or microbial growth functions and fit experimental parameters simultaneously to learning the dynamical system.

[1] Che, Zhengping, et al. 'Recurrent neural networks for multivariate time series with missing values.' *Scientific reports* 8.1 (2018): 1-12.

[2] Dimri, V. P., and M. Ravi Prakash. 'Scaling of power spectrum of extinction events in the fossil record.' *Earth and Planetary Science Letters* 186.3-4 (2001): 363-370.

[3] Rico-Martinez, R., et al. 'Discrete-vs. continuous-time nonlinear signal processing of Cu electrodissolution data.' *Chemical Engineering Communications* 118.1 (1992): 25-48.

350c Intersecting Quantum Computing and Control with Materials

Jihan Abou Halloun, Keshav Kasturi Rangan, Henrique Oyama, Helen Durand

Recent years have seen significant advances in the manner in which materials are handled in process systems engineering, with efforts integrating process systems tools with materials discovery. For example, computer-aided molecular design methods [1], simultaneous product and process design [2], optimization-based materials design [3], and data science approaches [4] have been techniques used for moving toward more systematic computational materials creation. Despite the greater emphasis on materials design, there has not been much discussion in the process systems engineering community of materials control. However, modulation of various phenomena can induce changes in materials. For example, [5] explores using a laser to impact material optical properties. Works such as this indicate that a deeper exploration of materials control in the process systems engineering field may prove valuable within the context of the current efforts in materials engineering.

Motivated by these considerations, this work focuses on two topics related to the design and control of materials. In the first part of the talk, we discuss how the complexities of attempting to optimally design materials motivate an exploration of whether quantum computing might be useful. Specifically, inspired by [7], which uses machine learning to attempt to locate quantum circuits which give a desired result, we discuss the challenges of trying to figure out the structure of a hybrid quantum/classical algorithm that should be used for searching for a means to represent the relationship between some material properties and the ground state energy if this is not based on a pre-conceived notion based on physics [6]. For example, if it is desired to use a small number of qubits due to the fact that only one- and five-qubit devices are currently available for free from IBM's Quantum Experience, one could ask what might be achieved on a single-qubit device. In this case, measurement of the qubits at the end of the algorithm will give a 0 or a 1, with some probability. The series of gates can modulate the probability to some extent, but only a finite set of gates is available. Using a hybrid quantum/classical algorithm, where the probabilities are then related to the ground state energy through some function, provides more tuning knobs, but also greater challenges in attempting to locate a function to use in relating the probabilities to the ground state energy when there is not a motivating physical relationship. Furthermore, some means for relating the chemistry to the algorithm steps must also be incorporated. We also discuss how noise in the quantum devices can impact the difficulty of attempting to design useful strategies via this ad hoc approach. The second part of the talk will provide a discussion of control of materials from a process systems engineering viewpoint, particularly in the context of simulation of closed-loop control of materials, discussion of materials dynamics, and discussion of the types of inputs that might be used in modulating a material property.

[1] Samudra, A. P., andamp; Sahinidis, N. V. (2013). Optimization-based framework for computer-aided molecular design. *AIChE Journal*, *59*(10), 3686-3701.

[2] Eden, M. R., Jørgensen, S. B., Gani, R., andamp; El-Halwagi, M. M. (2004). A novel framework for simultaneous separation process and product design. *Chemical Engineering and Processing: Process Intensification*, *43*(5), 595-608.

[3] Hanselman, C. L., andamp; Gounaris, C. E. (2016). A mathematical optimization framework for the design of nanopatterned surfaces. *AIChE Journal*, *62*(9), 3250-3263.

[4] Eugene, E. A., Phillip, W. A., andamp; Dowling, A. W. (2019). Data science-enabled molecular-to-systems engineering for sustainable water treatment. *Current Opinion in Chemical Engineering*,

26, 122-130.

[5] Khalsa, G., Benedek, N.A. and Moses, J., 2021. Ultrafast control of material optical properties via the infrared resonant Raman effect. *Physical Review X, 11*(2), p.021067.

[6] Rangan, K. K., J. Abou Halloun, H. Oyama, I. Azali Assoumani, N. Jairazbhoy, H. Durand, and S. K. Ng, "Design and Control Resilience/Robustness: Relationships to Quantum Computing and Cybersecurity," *Proceedings of the IFAC Symposium on Dynamics and Control of Process Systems, including Biosystems (DYCOPS)*, Busan, Republic of Korea, in press.

[7] Cincio, L., Subaşı, Y., Sornborger, A. T., andamp; Coles, P. J. (2018). Learning the quantum algorithm for state overlap. *New Journal of Physics*, *20*(11), 113022.

350d Generalizing Physics-Informed Neural Networks to Unseen Boundary and Initial Conditions

Zachary Kilwein, Fani Boukouvala

Physics-informed Neural Networks (PINNs) have emerged as a competitive approach to solving forward and inverse differential equation problems across scientific and engineering domains[1]. Several works have shown how PINNs may be advantageous for solving Partial Differential Equation systems (PDE) compared to traditional methods, such as compatibility with high performance computing architectures, combining of data and mechanistic equations, and high flexibility with problem geometries [2]. However, a trained PINN only represents a PDE solution for a specific realization of boundary and initial conditions. If these change, the PINN must be retrained, and this is computationally expensive and limits applications of PINNs that may require faster predictions. In this work, we propose applying the universal approximation theorem to this modeling paradigm in order to train generalized PINNs that can accurately predict solutions when given unseen boundary and/or initial conditions. While a-priori training costs and model complexity increase, generalized PINNs must only be trained once and the resultant model can provide fast approximations to PDE solutions over the full problem domain. We will that this approach can enable PINNs to be employed in time-sensitive optimization problems, such as model predictive control and PDE constrained optimization, and compare our approach with conventional techniques in terms of both accuracy and computational cost.

References:

[1] Maziar Raissi, Paris Perdikaris, and George E Karniadakis. Physics-informed neural networks: A deep learning framework for solving forward and inverse problems involving nonlinear partial differential equations. Journal of Computational Physics, 378:686-707, 2019.

[2] Shengze Cai, Zhiping Mao, Zhicheng Wang, Minglang Yin and George E Karniadakis. Physics-informed neural networks for fluid mechanics: A review. Acta Mechanica Sinica, pg 1-12, 2022.

350e Use of Computational Fluid Dynamics (CFD) Modelling to Optimize Continuous Mixer Design

Chang Kai Wu, Marc Thibaut

The manufacturing of silicone polymer products involves many mixing steps of viscous fluids. It is a challenge to be able to predict mixing patterns, dead zones, and residence time distribution for different equipment designs. Mixing in silicone polymerization processes is an extremely important step to achieve desired characteristics, such as molecular weight distribution and viscosity. Inefficient mixing leads to unwanted properties; in some cases, significant fouling on the walls of the mixer can have a very negative impact on the mixing efficiency.

In this study, a continuous mixer is developed to mix the crosslinker with silicone polymer before the next step polymerization. The objective is to create a Computational Fluid Dynamics (CFD) model to predict mixing pattern, dead zone and residence time distribution. The developed model is used to design the mixer with reduced fouling of the mixer and accumulation of cured polymer. Proposed designs are validated through experimental studies.

350f Verification of Neural-Network-Based Explicit Control Systems Using Mixed-Integer Programming

Calvin Tsay, Jan Kronqvist, Alexander Thebelt, Ruth Misener

While recent decades have seen the development and widespread adoption of optimization-based control systems, notably model predictive control (MPC) [1], solution of the involved optimization problems in real time can be intractable, owing to model size, nonlinearity, etc [2]. Therefore, the MPC controller is often approximated using surrogate models; neural networks using rectified linear (ReLU) activation functions have gained particular attention, as they can accurately represent complex piecewise-affine functions [3-4]. In general, neural networks can serve as a form of explicit MPC by training them on MPC controllers offline, then using the trained model as a feedback controller online.

Although neural networks can accurately approximate complex functions and be evaluated quickly, they may also be prone to overfitting, limiting their adoption in risk-critical applications. To this end, several works have sought to analyze properties of neural network controllers, such as constraint satisfaction and stability [4-6] or Lyapunov stability [7].

In this work, we show how embedding ReLU neural network controllers in mixed-integer optimization formulations, e.g., see [8-9], enables analyzing extreme behavior of closed-loop system dynamics prior to deployment. We propose two optimization formulations to verify the "trustworthiness" of a neural network controller: (i) identifying the maximum deviation from the original control system and (ii) computing extreme values of system states. The proposed formulations are computationally demonstrated using practically motivated cases studies.

References:

[1] Simkoff, J. M., Lejarza, F., Kelley, M. T., Tsay, C., andamp; Baldea, M. (2020). Process control and energy efficiency. Annual Review of Chemical and Biomolecular Engineering, 11, 423-445.

[2] Vaupel, Y., Hamacher, N. C., Caspari, A., Mhamdi, A., Kevrekidis, I. G., andamp; Mitsos, A. (2020). Accelerating nonlinear model predictive control through machine learning. Journal of Process Control, 92, 261-270.

[3] Kumar, P., Rawlings, J. B., andamp; Wright, S. J. (2021). Industrial, large-scale model predictive control with structured neural networks. Computers andamp; Chemical Engineering, 150, 107291.

[4] Paulson, J. A., andamp; Mesbah, A. (2020). Approximate closed-loop robust model predictive control with guaranteed stability and constraint satisfaction. IEEE Control Systems Letters, 4(3), 719-724.

[5] Hu, H., Fazlyab, M., Morari, M., andamp; Pappas, G. J. (2020, December). Reach-sdp: Reachability analysis of closed-loop systems with neural network controllers via semidefinite programming. In 2020 59th IEEE Conference on Decision and Control (CDC) (pp. 5929-5934). IEEE.

[6] Karg, B., andamp; Lucia, S. (2020, December). Stability and feasibility of neural network-based controllers via output range analysis. In 2020 59th IEEE Conference on Decision and Control (CDC) (pp. 4947-4954). IEEE.

[7] Dai, H., Landry, B., Yang, L., Pavone, M., andamp; Tedrake, R. (2021). Lyapunov-stable neural-network control. arXiv preprint arXiv:2109.14152.

[8] Tsay, C., Kronqvist, J., Thebelt, A., andamp; Misener, R. (2021). Partition-based formulations for mixed-integer optimization of trained ReLU neural networks. Advances in Neural Information Processing Systems, 34.

[9] Anderson, R., Huchette, J., Ma, W., Tjandraatmadja, C., andamp; Vielma, J. P. (2020). Strong mixed-integer programming formulations for trained neural networks. Mathematical Programming, 183(1), 3-39.

351a Fabrication of Hierarchical Sorbents By a Combined 3D-Printing and in-Situ Phase Separation Process from Carbon Nanotube-Enriched Polymer Solutions

Jialing Xu, Cheryl Slykas, Adam Braegelman, Kevin Gabriel Alvarez, Thomas Kasl, Matthew Webber, Vivek Sharma, Bryan Boudouris, William Phillip

Sorbent platforms with structures tailored at the molecular through device scales are needed to meet the demand for separation systems that can isolate trace analytes from complex mixtures in an energy-efficient manner. Non-solvent induced phase separation is a common technology used to create thin polymer membrane-based sorbents, which are attractive due to their rapid mass transfer characteristics. However, membrane sorbents suffer from relatively low binding capacities. While smaller pores and thicker membranes can increase the surface area and number of binding solute sites, these changes will inevitably decrease the permeability of the membranes. Addressing this tradeoff and translating lab scale discoveries of binding chemistries to functioning devices requires the development of manufacturing methods that provide control over sorbent structures at the molecular through meter scales. Here, direct-ink-writing (DIW) 3D printing is combined with a surface-segregation and vapor-induced phase separation (SVIPS) process to create hierarchically-structured sorbents that satisfy this criterion.

Composite inks containing polysulfone, polystyrene-*block*-poly(acrylic acid) (PS-PAA), and carbon nanotubes were extruded to form woodpile structures with micrometer-scale channels between the printed filaments. Simultaneously, because the DIW process occurred within a controlled humidity environment, the SVIPS process generates a fully-interconnected network of nanoscale pores within the filaments as they are extruded to form the sorbent matrix. The ink formulation is critical to enabling the combined DIW and SVIPS method. For example, shear and Dripping-onto-substrate (DOS) rheometry measurements were used to demonstrate the influence of carbon nanotubes on the shear and extensional viscosities of the ink. Specifically, adding carbon nanotubes resulted in more pronounced shear thinning behavior, increased the extensional viscosity, and delayed the pinch-off time, which enable the deposition of more regular micrometer-scale patterns. Additionally, quantifying the thermodynamic properties of the ink demonstrates their impact on the gelation time scale within different humidity environments. The ability to engineer these natural time scales through the ink formulation and modify the printing conditions to match the processing time with these time scales provides insights for designing protocols that can be extended to different ink systems, as well as providing for better control over the hierarchical sorbent structures.

In addition to creating high surface area nanoporosity, the SVIPS mechanism drives the PAA blocks to the surface of the pore walls, thereby providing functional handles that allow the sorbent chemistry to be further tailored. In this work, sequential carbodiimide coupling reactions were used to introduce branched-polyethylenimine (PEI) and then terpyridine (Terp) functionality, which increased the density of binding sites and affinity towards transition metal ions, respectively. The metal ion capture performance that results from this molecular design is demonstrated through the selective recovery of Co^{2+} from solutions containing cobalt and lithium dissolved at pH 1. At the device-scale, the microstructure of the sorbents results in high throughput devices (i.e., hydraulic permeabilities of $\sim 10^5$ L m^{-2} h^{-1} bar^{-1}) that demonstrate the efficient (>95%) removal of copper from 1 ppm feed solutions during dynamic flowthrough experiments.

In conclusion, incorporating 3D printing technique with SVIPS casting enables the creation of sorbents with well-controlled hierarchical structures that achieve high binding capacity, affinity, and permeability simultaneously. Ultimately, the successful design, printing, and control of porous nanostructure provide new strategies for the fabrication of next-generation sorbents.

351b 3D Printing of a Recycled Terephthalic Acid-Based Copolyester Containing Tetramethylcyclobutanediol

Samarthya Bhagia, Surbhi Kore, Sanjita Wasti, Jaroslav Durkovic, Jan Kovac, Xianhui Zhao, Uday Vaidya, Soydan Ozcan, Arthur Ragauskas

Fused deposition modeling (FDM) is an important class of additive manufacturing that offers scalability and thermoforming of complex designs using thermoplastic matrices. A terephthalic acid-based polyester containing tetramethylcyclobutanediol (TMCD) diol monomer[3] was recycled by 3D printing for the first time. The rigidity of cyclobutane ring in TMCD provides higher impact strength and glass transition temperature compared to other terephthalic acid copolyesters. The printing processes were optimized and materials were characterized for mechanical strength, thermal behavior, morphology and polymer structure. We found that bed temperature near ~100 °C was important for adhesion of first layer to the printing bed. The tensile strength, Young's modulus and notched Izod impact strength of the printed sample were 44 MPa, 4.2 GPa and 16.4 kJ/m^2, respectively. Glass transition temperature of the copolyester was 103 °C. Weight-average molecular weight decreased from 38 to 32 kDa on recycling by 3D printing. ^{1}H, ^{13}C, ^{1}H-^{1}H COSY and ^{1}H-^{13}C HSQC NMR showed that fine structure was not altered by high temperature in 3D printing. We also performed micro-computed tomography (μCT) to see the voids among roads and layers in printed composites. The utilization of this copolyester for making 3D printed parts can increase recycling and the printed parts have properties that are suitable for making commercial products.

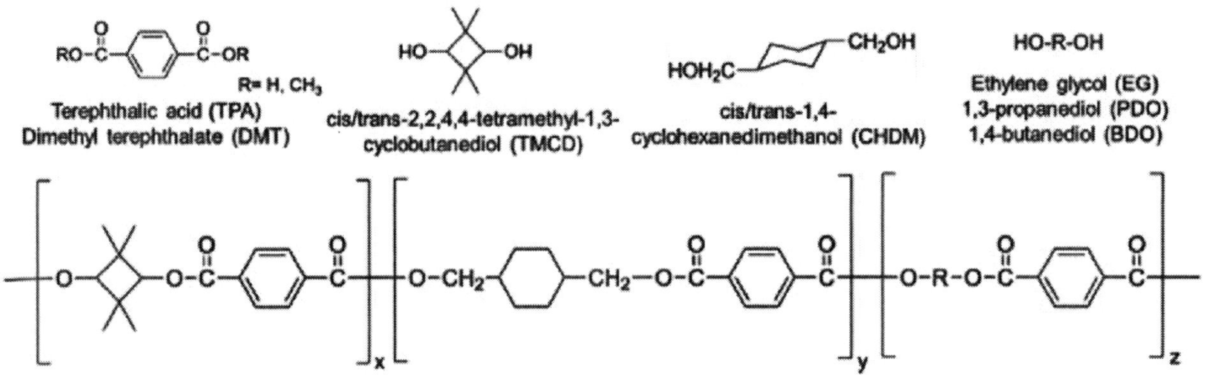

351c Single-Digit Micron CLIP 3D Printing, Modeling and Applications

Kaiwen Hsiao, Gabriel Lipkowitz, Eric Shaqfeh, Joseph M DeSimone

The current landscape of high-resolution 3D printing technologies typically involves a tradeoff between scalability and resolution. Existing technologies have achieved nanometer scale print precision but suffer from much lower print speed, making the technology not easily scalable. Here, we present a novel 3D printing technology that combines micro-DLP with Continuous Liquid Interface Production (CLIP) technology that can print at single-digit micron scale resolution with a print speed up to 200 times greater than other existing high-resolution printing technologies.

The design and printing strategy of the single-digit micron CLIP printer involves several sub-components. To achieve single-digit micron resolution, we have designed and implemented a home-built projection optics system, an in-line camera system to monitor the projection pattern and a stand-alone contrast-based focusing algorithm to achieve an optimal projection plane. To optimize and provide a full understanding of the high-resolution CLIP 3D printing process, we have developed a physical model that involves Zemax based optics point spread function prediction, reaction kinetics and momentum transport model using lubrication theory that provides insights into the photopolymerization gradient, flow velocity profile and mass transport in the CLIP system for both Newtonian and non-Newtonian polymeric materials. It is observed both experimentally and from modeling that a delicate control of UV projection intensity, printing step size, and waiting time between exposures are critical to achieving the desired resolution. Finally, a physical model informed software-controlled printing strategy is adopted. The step-and-expose printing process allows us to resolve single-digit micron prints with high-precision.

The high-resolution CLIP platform has shown its potential to print with various materials (including hydrogels and elastomeric materials), with applications ranging from drug delivery and continuous health monitoring platforms to microfluidic device fabrication, force sensors and electronic packaging devices. We envision this new technology can be transformative in providing researchers across many domains with improved access and freedom to form and create designs.

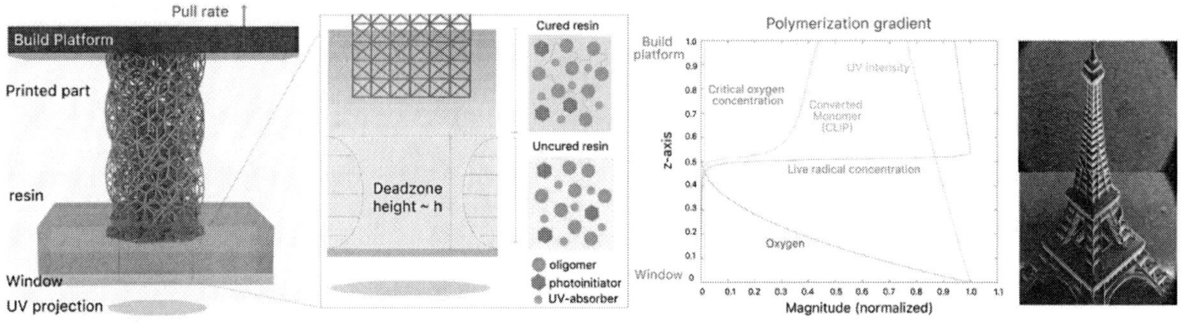

351d Additive Manufacturing Using Olefin Metathesis

Samuel Leguizamon, Adam Cook, Leah Appelhans, Jeffrey Foster

The additive manufacturing (AM) of thermoset polymers has been limited primarily to radical or cationic photopolymerizations, restricting available resins and attainable material properties. Ring-opening metathesis polymerization (ROMP) has emerged as a useful mechanism for the printing of thermoset materials with advantageous high-performance properties. However, previous demonstrations of ROMP for AM have been hindered by sluggish printing rates and restricted to a small number of patterned layers or simple geometries. Here, we present the use of photosensitizers and latent ruthenium catalysts to accelerate the photo-initiated ROMP of dicyclopentadiene resins for the 3D printing of tough, high-performance thermosets. Both UV-assisted direct-ink write (DIW) AM and stereolithography (SLA) AM are demonstrated. Notably, continuous SLA AM is achieved by employing a photo-base generator to yield a catalyst deactivation layer adjacent to the projection window. With the use of the resin system described, DIW and continuous-SLA printing rates of 60 $mm \cdot s^{-1}$ and 180 mm h^{-1} are demonstrated.

Sandia National Laboratories is a multimission laboratory managed and operated by National Technology and Engineering Solutions of Sandia, LLC, a wholly owned subsidiary of Honeywell International Inc., for the U.S. Department of Energy's National Nuclear Security Administration under contract DE-NA0003525.

351e Pressure-Based Process Monitoring of Direct-Ink Write Additive Manufacturing

Jessica Kopatz, Derek Reinholtz, Jonathan Leonard, Alexander Tappan, Adam Cook, Anne M. Grillet

Additive manufacturing allows for rapid prototyping, the creation of unique and complex geometries, the generation of less material waste, among many other benefits, when compared to traditional manufacturing. Direct-ink write is an extrusion-based additive manufacturing technique that enables the creation of custom-designed and reproducible hardware via extrusion of a paste or ink. Therefore, inherent changes in ink rheology and inhomogeneous ink formulation, due to insufficient mixing, variable pot-life, etc., can result in non-uniform extrusion and cause defects that are detrimental to the properties of the printed artifact. Additionally, inks containing a high particle loading and using particles with complex morphologies can contain trapped air or particle aggregates that cause undesirable clogging during extrusion, both of which can affect printed part properties and quality. The integration of pressure sensors on the extrusion apparatus can monitor the print process in situ by identifying changes in flow during printing. The goal of this work is to identify the viability and sensitivity of pressure-based process monitoring during extrusion to detect printing defects. Pressure is measured both at the top of the syringe barrel and just before the nozzle during extrusion of inks. To determine the sensitivity of the pressure measurement to aggregates in the ink, engineered defects in the form of 100-400 micron spherical polymer particles are added to DowSil SE1700 commercial printing ink resin. Pressure measurements over a range of extrusion conditions and engineered defect sizes as well as image analysis of the printed bead shape are used to determine sensitivity thresholds for the process monitoring.

SNL is managed and operated by NTESS under DOE NNSA contract DE-NA0003525

351f Versatile Additive Manufacturing of Microscale Metals and Alloys Via Hydrogel Infusion

Max Saccone, Daryl Yee, Rebecca Gallivan, Kai Narita, Julia R. Greer

Additive manufacturing (AM) of metal components has emerged as a uniquely powerful tool for rapid prototyping and creating parts with complex geometries and applications from aerospace to biomedical industries. We explore AM enabled by digital light processing (DLP), typically used to produce polymers and ceramics, for the fabrication of metals and metal alloys. This addresses a long-standing gap in conventional metal AM processes, which struggle to fabricate metals at the mesoscale, with feature sizes in the 10-100 μm range.

We developed a general method for the fabrication of a wide variety of metals and alloys with complex shapes, mesoscale resolution, and overall cm-scale dimensions via DLP AM and subsequent post-processing and thermal treatment. This streamlined technique makes use of a single resin composition and a single set of processing conditions during the DLP process, followed by infusion of appropriate metal precursors into a hydrogel structure. Heat treatment in oxidizing followed by reducing atmospheres converts the polymer/precursor matrix into the target metal. Unlike previous vat photopolymerization strategies which have target materials or precursors incorporated into the resin during the printing process, this method does not require re-optimization of resins and resin curing parameters when the target material is changed; relevant process control parameters are shifted to the hydrogel infusion and heat treatment steps.

We report photoresin design and post-processing strategies to show how to use this technique to easily fabricate a range of AM metal and alloy structures. As a proof of concept, we fabricated octet lattice architectures with beam diameters on the order of 50 μm from several materials including copper, nickel, silver, cobalt, cupronickel alloys, tungsten, and multicomponent high entropy alloys. Furthermore, we demonstrate fabrication of multi-material metal structures via this method. Throughout, we investigate the relationship between processing and microstructure of these additively manufactured materials and highlight the wide-ranging versatility of achievable materials and microstructures.

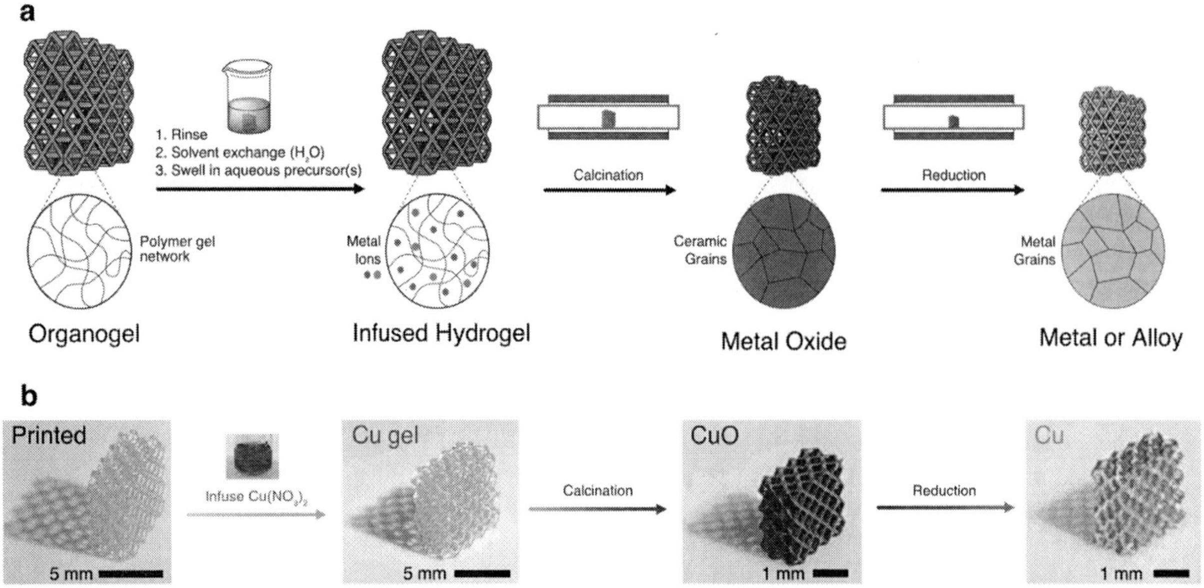

Figure 1 | Schematic of hydrogel infusion additive manufacturing. a, A 3D printed organogel structure is converted to a hydrogel structure and infused with metal salt precursors. Subsequent calcination and reduction transform the infused hydrogel to a metal or alloy. **b,** Example of hydrogel infusion additive manufacturing of a copper octet microlattice with 40 μm beam diameter.

156b Keynote Talk-Machine Learning Based Software Tool for Efficient Wastewater Asset Management

Kirti Yenkie, Jake Stengel, Emmanuel Aboagye, Matt DeNafo

As per the statistics in 2017[1], more than 27 million Americans are exposed to water systems in violation of the health standards established in the Safe Drinking Water Act (SDWA).[2] The American Society of Civil Engineers (ASCE) infrastructure rating systems have rated the Unites States' Water (W) and Wastewater (WW) category as D and D+ respectively[3], which is alarming and directly correlates to the issues of Water Quality, Equity, and Public Health. Thus, there is an immediate need for convergent thinking and holistic solutions for enhanced Wandamp;WW management systems that concurrently address infrastructure improvements, as well as enable significant revisions in the water and wastewater quality monitoring, treatment, and management systems, to meet the SDWA guidelines.

Most water utility companies use a reactionary or subjective approach[4] for safeguarding water quality and infrastructure. The problem with a reactionary (wait-watch-act) approach is that problems usually cannot be solved until a failure or hazardous event occurs. Once an issue arises, significant amounts of manpower and capital are needed to access and repair the infrastructure failures[5], and, at the same time, it is important to minimize any negative impacts on the surrounding environment and the quality of water in the distribution systems. Thus, the reactionary approach leads to large capital losses and negative environmental impacts due to the lack of a backup system or plan of action. The new enabling ML-based software tool developed by our team will serve municipalities, utility companies, and facilities for informed decision-making and timely management of their wastewater infrastructure. We have combined methods from data collection, analysis, mathematical modeling, artificial intelligence, machine learning[6], and optimization[7] to develop the software tool which will perform vulnerability and risk assessment of existing utilities and infrastructure and create plans of action towards resilience and manage the wastewater collection systems effectively.

References:

(1) PolicyLink. *Water, Health, and Equity - Clean Water for All*; 2017; pp 1–32.

(2) US EPA, O. *Safe Drinking Water Act (SDWA)*. US EPA. https://www.epa.gov/sdwa (accessed 2021-02-08).

(3) America's Infrastructure Grade. *ASCE's 2017 Infrastructure Report Card.*

(4) Barles, R. Drinking Water Infrastructure Needs Survey and Assessment. 76.

(5) *Aging water infrastructure 'nearing the end of its useful life' - Investigate MidwestInvestigate Midwest.* https://investigatemidwest.org/2014/06/23/aging-water-infrastructure-nearing-the-end-of-its-useful-life/ (accessed 2021-02-09).

(6) Kumar, A.; Flores-Cerrillo, J. Machine Learning in Python for Process Systems Engineering. 352.

(7) Roshani E.; Filion Y. R. Event-Based Approach to Optimize the Timing of Water Main Rehabilitation with Asset Management Strategies. *Journal of Water Resources Planning and Management* **2014**, *140*, 04014004. https://doi.org/10.1061/(ASCE)WR.1943-5452.0000392.

91b Partially-Connected Recurrent Neural Network Modeling for Predictive Control Using Noisy Data

Mohammed Alhajeri, Zhe Wu, Panagiotis Christofides

Over the last four decades, neural networks have demonstrated their ability to provide adequate modeling approximations of complex, highly nonlinear, and stochastic chemical processes when utilized as machine learning (ML) techniques for control and state prediction of nonlinear processes [1,2,3]. Recurrent neural networks (RNN) have been widely used for modeling a broad class of dynamical systems with objectives of control and state estimation. In our previous work [4], a model predictive control (MPC) that uses a process structure-based RNN model, termed partially-connected RNN, was proposed and achieved improvements in closed-loop performance and computational time relatively to the standard (i.e. fully-connected) RNN models. However, in the development of the partially connected RNN models it is assumed that both the given system and the data are noise-free. In fact, state measurements in a variety of chemical industrial operations are corrupted by noise. Therefore, it is necessary to study the effect of noise when utilizing the underlying modeling approach.

This work investigates the effect of both gaussian and non-gaussian noisy data on the performance of partially-connected RNN model approximation of class on multi-input-multi-outputs systems. Furthermore, two different RNN building methods, specifically Monte Carlo dropout rate and co-teaching methods, are utilized in development of partially connected RNN-LSTM. These two techniques are employed to improve the open-loop accuracy as well as the closed loop performance under Lyapunov based MPC. Aspen Plus Dynamics, which is a well-known high-fidelity software, is used to develop a large-scale chemical process example in order to demonstrate the anticipated improvements in both open-loop approximation and controller performance in two cases: gaussian noise, and non-gaussian noise. In the presence of each noise type, both open-loop and closed-loop simulation will be carried out, analyzed, and discussed.

References:

[1] Sontag, E. D. 'Neural nets as systems models and controllers.' Proc. Seventh Yale Workshop on Adaptive and Learning Systems, 1992.

[2] Mohanty, Swati. 'Artificial neural network based system identification and model predictive control of a flotation column.' Journal of Process Control 19.6 (2009): 991-999.

[3] Wu, Z., A. Tran, D. Rincon and P. D. Christofides, 'Machine Learning-Based Predictive Control of Nonlinear Processes. Part I: Theory," AIChE J., 65, e16729, 2019

[4] Alhajeri, M., J. Luo, Z. Wu, F. Albalawi and P. D. Christofides, 'Process Structure-Based Recurrent Neural Network Modeling for Predictive Control: A Comparative Study," Chem. Eng. Res. andamp; Des., 179, 77-89, 2022.

91c Handling Correlated Data for Artificial Neural Network (ANN)-Based Model Predictive Control (MPC) Implementations

Hesam Hassanpour, Brandon Corbett, Prashant Mhaskar

Advanced control systems play a key role in chemical process industries to achieve economic and safety objectives. However, there have been several challenging issues in developing these systems such as nonlinearity, process constraints, and multi-variable interactions. Model predictive control (MPC) is one of the optimal control techniques that can successfully handle these challenges [1]. Using an optimization problem, MPC decides which control action to implement by testing the effect of a candidate input on the process using a process model. Thus, good control performance relies on using an accurate process model. Several modeling techniques including first-principles and data-driven based approaches have been developed and used for MPC implementations. Although first-principles models have good extrapolation capabilities, they are generally difficult to develop and maintain because of uncertainty in physical parameters and unmeasured states.

With the recent development of computing and data storage technologies, there has been increasing interest in building data-driven and machine learning-based models. Among many machine learning techniques, artificial neural networks (ANNs) have received significant attention due to their ability to capture nonlinearity. ANN-based models have also been used to implement MPC [2,3]. To develop an ANN-based model, a significant amount of high-quality data is usually required because of a large number of parameters that need to be tuned. However, the presence of redundant information in some historical data can exacerbate the negative impact of the over-fitting problem. The redundancy issue usually happens when data are collected from a process operating under closed-loop conditions. In this situation, if data samples are directly used (without utilizing an appropriate data mining mechanism) to develop MPC, it may lead to a poor closed-loop performance. Therefore, additional constraints must be considered within the standard ANN-based MPC framework to maintain model validity [4].

Motivated by the above considerations, this work addresses the problem of ANN-based MPC design, where the collected closed-loop data contains correlation in both the input and output spaces. To maintain model validity, a principal component analysis (PCA)-based squared prediction error (SPE) constraint is first added to the standard ANN-based MPC formulation to make the control actions follow the same correlation as that of the training input data (constrained ANN-based MPC). Note that implementing the ANN-based MPC just by adding the PCA-based SPE constraint on the input moves does not necessarily guarantee that the outputs will track the arbitrarily prescribed set-points because of the existing correlation in the training output data. To overcome this problem, an optimization problem is used to minimize the sum of squared error between the prescribed and achievable set-points by applying a PCA-based SPE constraint, developed using the training output data. In addition, to handle the nonlinear correlation between the process outputs, an autoencoder-based optimization problem is developed to calculate the achievable set-points. The effectiveness of the proposed approaches is shown via simulation of a chemical reactor example. The results reveal the superior performance of the constrained ANN-based MPC over the standard ANN-based MPC to drive the outputs to the achievable set-points. Further improvement in closed loop performance is achieved using the achievable set-points calculated based on the autoencoder-based strategy (due to the superior performance of the autoencoder to handle the nonlinear correlation, while calculating the achievable set-points) [5].

[1] Mayne, D. Q., Rawlings, J. B., Rao, C. V. andamp; Scokaert, P. O. (2000). Constrained model

predictive control: Stability and optimality. *Automatica, 36*(6), 789–814.

[2] Sadeghassadi, M., Macnab, C. J., Gopaluni, B., andamp; Westwick, D. (2018). Application of neural networks for optimal-setpoint design and MPC control in biological wastewater treatment. *Computers andamp; Chemical Engineering, 115,* 150-160.

[3] Wu, Z., Rincon, D., Luo, J., andamp; Christofides, P. D. (2021). Machine learning modeling and predictive control of nonlinear processes using noisy data. *AIChE Journal, 67*(4), e17164.

[4] Hassanpour, H., Corbett, B., andamp; Mhaskar, P. (2020). Integrating dynamic neural network models with principal component analysis for adaptive model predictive control. *Chemical Engineering Research and Design*, *161*, 26-37.

[5] Hassanpour, H., Corbett, B., andamp; Mhaskar, P. (2022). Artificial Neural Network-Based Model Predictive Control Using Correlated Data. *Industrial andamp; Engineering Chemistry Research.*

91d Application of Data Driven Techniques to Industrial Hydroprocessing Units

Debanjan Ghosh, Jesus Moreira, Prashant Mhaskar

Hydoprocessing units in petroleum refineries play a critical role in removing impurities from the crude and crack the heavier crude to form lighter products for subsequent operations. These units have an intricate network of several unit operations and modeling such units play a pivotal role in estimating/predicting values of important variables, improving the control and optimization of the plant for efficient operation among several other applications. This presentation demonstrates the development and implementation of linear dynamic data-based models in estimating product qualities and other key monitoring variables in the hydroprocessing unit of an industrial refinery. Historical operational data from two different units was used and appropriate data-driven modeling strategies were formulated to address this problem. The usefulness of Dynamic-Partial Least Squares (DPLS) [1] and Subspace Identification [2] in the estimation of important variables of the units is demonstrated. In this study, the Subspace Identification methodology was suitably adapted and made more generalizable to handle missing data in both inputs and outputs. The Non-linear Partial Least Squares (NIPALS) algorithm, being robust to missing data, was used to replace some of the conventional steps in the Subspace Identification. The identified linear time invariant (LTI) model was then embedded inside a Kalman filtering state estimation technique to make use of the current process measurements to correct the model states as the process keeps marching along in time. The methods used in this study not only elegantly handle issues like missing data problem associated with real data sets but also conveniently address the problem of overfitting associated with data models, and the results demonstrate the suitability of using these linear dynamic models for specific problems in the context of hydroprocessing refinery. The results are a significant improvement over the current modeling techniques used in the industry where the units are operated and suggests practitioners the usefulness and potential application of these models.

[1] Wenfu Ku, Robert H. Storer, Christos Georgakis, Disturbance detection and isolation by dynamic principal component analysis, Chemometrics and Intelligent Laboratory Systems, Volume 30, Issue 1, 1995, Pages 179-196, ISSN 0169-7439

[2] Patel, N, Mhaskar, P, Corbett, B. Subspace based model identification for missing data. *AIChE J.* 2020; 66:e16538.

2022 AIChE Annual Meeting

INVESTIGATING MACHINE LEARNING TECHNIQUES FOR EFFECTIVE PREDICTIVE MAINTENANCE IN INDUSTRIAL SYSTEMS

Mahdi Ahmed, Louis Allen, and Joan Cordiner, University of Sheffield, Sheffield, United Kingdom

I. Abstract

Predictive maintenance (PdM); this is predicting the time until a fault of unit operations in an industrial process, identifying what these faults are, and their root causes in advance. Using this information can enable maintenance scheduling to be optimized, process safety to be improved, and a reduction in downtime to over conventional preventative strategies. Nonetheless, additional research is required to extend current predictive strategies that only analyze single-unit operations, failing to consider the underlying links between sequential unit operations of an entire industrial process. In this study, the objective was to use different Machine Learning (ML) techniques to forecast the future behavior of sensor data by taking a multivariate approach and incorporate these relationships between multiple unit operations. We aimed to improve forecasting accuracy by leveraging this additional data. This was accomplished using data from a continuous powder to tablet processing pilot plant, composed of 9-unit operations, and equipped with a total of 47 sensors, mimicking the scale of industrial systems. The effectiveness of different ML models was evaluated in two scenarios; short-term, for forecasting 2 hours into the future, and long-term, where degradation was simulated and forecasted over a 90-day longer period. An LSTM (Long-Short-Term-Memory Neural Network) model produced the most accurate forecasting results, predicting varying stochastic patterns of degradation over 73 days in advance, whilst also providing sufficient detail that enabled classification. The results also demonstrated the value of taking a multivariate analysis throughout the entire process, thereby improving forecasting accuracy significantly compared to a univariate approach. In contrast, more traditionally used linear and naive models fail to capture the complexities of the system, resulting in faults being predicted either too early or too late, corresponding to money being wasted on unnecessary maintenance. Using the present research, we develop the foundation for more sophisticated PdM effectively forecasting data which can be used to pre-emptively mediate faults in an industrial process.

II. Introduction

As of 2022, the PdM size was $4.5 billion dollars. By 2030, it is expected to be worth over $65 billion dollars [1]. The cost of downtime is substantial and using PdM allows for a fault of an industrial process to be predicted and characterized in advance so that it can be prevented from escalating into permanent damage. Costs associated with shutdowns and maintenance are greatly reduced, additionally improving sustainability and safety of the plant with the knowledge of both preparing for, and mitigating an anticipated fault, allowing for maintenance scheduling to be optimized. A preliminary study has already demonstrated that maintenance costs can be reduced up to 44% through optimizing maintenance scheduling alone [2]. This is typically achieved through monitoring and analyzing sensor data of an industrial system. Abded-Hameed discussed how industrial processes can typically degrade overtime, following stochastic mathematical patterns [3]. Once a process has degraded beyond a certain measurable point, this is defined as a fault or failure. Thus, the viability of PdM is evaluated, by implementing different ML algorithms, as they have previously been proven to be effective in analyzing and characterizing complex patterns and interactions in industrial data. This enables us to do two key things; classify, and forecast, these degradation patterns.

A classification method takes anomalous or conspicuous patterns of current data and maps this to pre-defined faults. Carvalho et al outlined several techniques, such as decision trees (Random Forests and Gradient Boosting), and neural networks, with strong performances in identifying these patterns [4].

Forecasting involves analyzing past/current data and predicting its future behavior. A degradation pattern can be extrapolated to establish the time taken for degradation to culminate into an eventual fault. There is, however, little application of advanced forecasting algorithms in industry, with present prognostic solutions relying on simple linear or naive models. These are well-known and easy to implement real-time for simpler, and specific industrial unit operations, whilst still providing "good enough" forecasts in this context. Carvalho et al confirmed this, noting a preference for units such as motors, compressors, pumps, and turbines, where a substantial amount of information about degradation can be obtained from using vibrational data. Forecasting algorithms practically used in industry tend to only tackle unit operations of this type, because of these discernible signs of degradation that simpler models can evaluate. Orhan et al. concluded however, that vibrational data is only helpful for very specific units, and hence why current PdM strategies typically focus on these listed unit operations [5]. The process of identifying degradation signs for units in contrast to this must be performed using other types of sensor data - which is far more challenging.

This has highlighted the need for algorithms that can tackle analyzing data from an entire industrial process – not just a single unit operation, as the links between different units are currently neglected in many cases, where an abundant amount of information is held. Access to additional information will allow us to detect and accurately model degradation in units that do not contain moving parts which are much harder to identify long-term degradation patterns in. Characterization of these links is also critical in developing methods for preventing faults from occurring through changing key process parameters.

Objectives

To address this key issue, this paper proposes to develop a model that incorporates information from sensors across an entire process operation (in contrast to the traditional approach of models focusing on single specific unit operations) to enable and improve degradation forecasting and characterization of sensor measurements. It is our goal to demonstrate that adding these missing links will enhance accuracy over a univariate approach and provide a framework for more comprehensive PdM solutions.

III. Case Study

The continuous powder to tablet processing pilot plant (DiPP) at The University of Sheffield consists of 9-unit operations (including crystallization, blending, granulation and tableting) containing 47 different sensors, mimicking the scale of industrial systems, and is thus extremely relevant for carrying out the proposed objective [6].

Data was collected in multiple runs (sensors are logged in-line every 5 seconds), yet in each run, there was no observable degradation or anomalies that resulted in failure or shutdown. As the plant is used for training and educational purposes, it operates from 1 hour to up to 3 hours at a time, which isn't sufficient time for significant changes to be observed in the equipment.

Hence, this dataset is presented in two variations. A "smaller" dataset, consisting of sensor readings over a period of two hours (*1,440* readings per sensor for a total of *67,000* data points). A "larger" dataset is also used, which was constructed by concatenating several runs up to 90 days, with sensor readings sampled every five minutes instead, to minimize computational load (*25,800* readings per sensor for a total of *1.2 million* data points). Degradation models (simulated degradation) were applied only to the larger set (expanded upon in the methodology section), first to keep the smaller set as "real" as possible to test shorter term forecasting (such as soft sensors and predictive control) whilst avoiding any bias, but also as a mechanism for evaluating longer-term degradation, and at the same time maintaining the natural noise and patterns of the system, minimizing the influence of using simulated data. Section IV will introduce the methodology in which this case study is applied to.

IV. Methodology

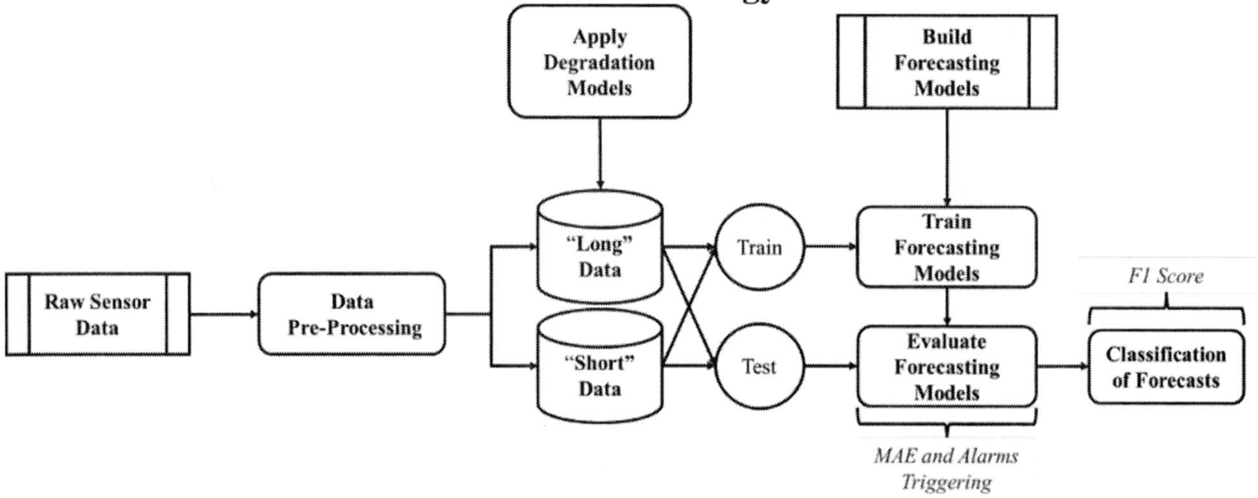

Figure 1. Data flow and methodology

Figure 1 represents a flow chart of the methodology, overviewing the process of building, training, and evaluating models on data that has been pre-processed and split into different scenarios.

In this study, VAR (Vector Autoregression), XGB (Gradient Boosting), and LSTM models were used since they have been previously identified as popular forecasting algorithms in finance. These models are capable of multivariate forecasting (using many features) across the dataset. Each model's forecast is assessed based on mean absolute error (*Equation 1*). For testing short-term patterns, a sensor measuring the outlet humidity of a fluid bed dryer was chosen. As it is an "outlet", it is not controlled directly, and is influenced by changes in other parameters of the system, such as flowrates and temperatures of upstream unit operations. Additionally, the sensor is vital for determining the moisture content of the tablets produced at the plant, which is then used to establish the tablet's physical properties, thus making it an important sensor to forecast. From this, one can predict the impact of parameter changes in the system, its impact on the overall properties of the tablets, and even use as a soft sensor or for predictive control.

Due to the lack of observable long-term degradation, degradation models were produced to match non-linear and stochastic patterns typically seen in processes as discussed by Abded-Hameed, whilst keeping as like a real industrial system as possible. These degradation models can therefore be applied to the dataset to simulate DiPP unit operations deteriorating in various ways over time.

The degradation patterns are made to mimic typical long-term failures of the plant, and sensor 9 (outlet pressure of a fluid bed dryer) was chosen for long-term forecasting. This must maintain a set-point, and any deviation from that value is indicative of unit fatigue over time. This would be a consequence of drying particularly sticky materials, leading to buildup of this on the spray head, and thus resulting in higher pressure drops (increasing volatility) and poorer fluidization of the outlet material (non-linear decline). After a certain degree of degradation, different alarms would trigger as a result. To evaluate the forecasting performance at a longer-term basis, these alarm points are selected based on the extent of degradation, and how well the model predicts the time the alarm would have triggered.

Equation 1: $$MAE = \frac{\sum_{i=1}^{n}|y_i - x_i|}{n}$$

Where $\mathbf{x_i}$ is the true value, $\mathbf{y_i}$ is the prediction, $\mathbf{n}$ is the total number of data points

V. Results

Short-Term Forecasting

Each forecasting model was first tested on a short-term basis, as shown in *Table 1*. The number of features used to train the model was also varied to observe its effect on each of the models – A model that is robust to the choice of features used is generally better. A univariate approach was used to measure the effect compared to a multivariate approach. Where "matrix features" are used, a causation matrix was generated (grangers causation matrix), which tests the "causality" between different features. Where "related features" is used, this is grouping the sensors the dataset using domain knowledge of the system, i.e., it is evident that the relative humidity of the dryer would have a cause relationship with the temperature and pressures of upstream unit operations, and so sensors can be grouped together with this in mind. Finally, "all features" is using every single feature that was available in the dataset. In each feature scenario, the LSTM performs the best for short-term forecasting.

Table 1. Comparing the accuracies (mean absolute error) of short-term forecasting of each model using differing features, forecasting 1.4 hours into the future, trained on the 0.54 hours of data

Features Used	**Model Accuracy (Mean Absolute Error)**			
	Linear	**VAR**	**XGB**	**LSTM**
Univariate	3.58	0.4701	9.831	0.2104
Related Features		0.2812	4.423	0.1026
Matrix Features		0.3611	4.484	0.1036
All Features		7.671	4.542	0.1032

Long-Term Forecasting

Figure 2 shows four ways in which degradation was simulated. A model's accuracy is measured by its ability to predict when a specified alarm point will trigger (i.e., when degradation is severe enough that it will result in a process fault), and how far off it is from predicting the actual time. Each model is trained on 18 days' worth of data, forecasting the next 72 days. The LSTM had the smallest error in each degradation scenario. This is summarized in *Table 2.*

Table 2. Comparing the accuracy of each ML model (number of days error) at forecasting when alarm 1 and 2 would trigger due to on-going degradation

Degradation Pattern	**Model Forecasting ability (No of days error)**					
	Alarm 1			**Alarm 2**		
	Linear	**VAR**	**LSTM**	**Linear**	**VAR**	**LSTM**
Sigmoidal Degradation	<u>0</u>	<u>0</u>	<u>0</u>	25	5	<u>0</u>
Inverse-Sigmoidal Degradation	-2	-4	<u>1</u>	-16	-19	<u>0</u>
Sigmoidal Degradation + Instability	10	<u>5</u>	<u>5</u>	17	13	<u>3</u>
Increasing Instability	-	-	<u>3</u>	-	-	<u>-3</u>

Classification

Given that the LSTM provided the best forecasts, classification models were trained on each degradation pattern. The output of the was LSTM fed into these to better examine the "shape" of the forecasts and see if they are detailed enough to also be classified. A degradation pattern that can also be classified, allows for appropriate recommendations to be made depending on the predicted fault, and a framework for automation of this. A sequenced classification LSTM model was for this purpose. The model's F1 scores (*equation 2*) for classifying each pattern is shown in *Table 3.* The F1 scores for each degradation pattern are strong, except for sigmoidal degradation.

$$\textbf{Equation 2: } F1 = \frac{2(P \times R)}{P+R}$$

Where P is precision and R is recall

Table 3. F1 score of classifying each of the forecasts made by the LSTM model, using a sequenced classification LSTM model

Degradation Pattern	**F1 Score**
Normal Operation	0.999
Sigmoidal Degradation	0.304
Inverse-Sigmoidal Degradation	0.740
Sigmoidal Degradation + Instability	0.964
Increasing Instability	0.760
Total Average	**<u>0.757</u>**

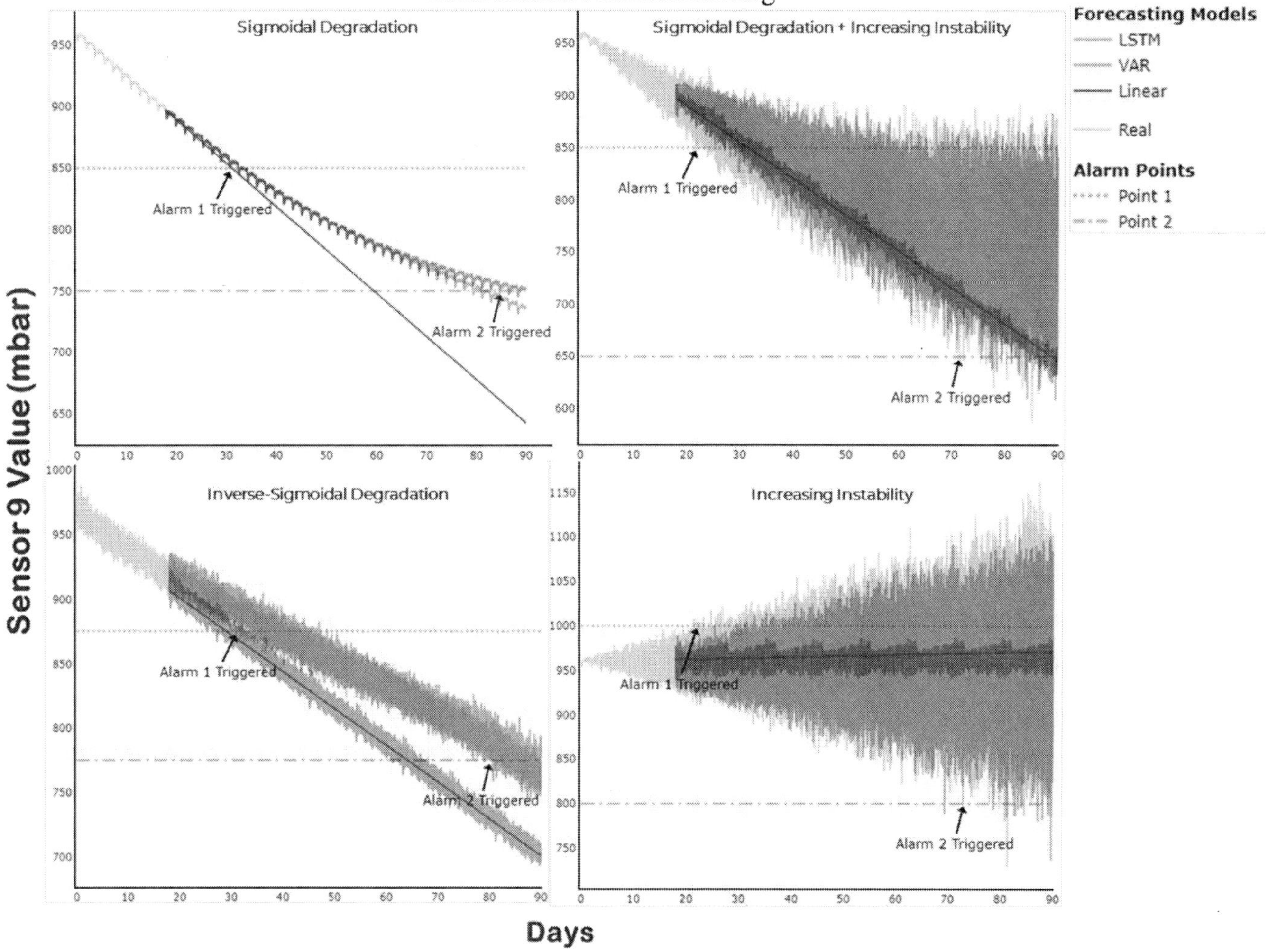

Figure 2. Forecasting performances of different ML models for 4 different degradation patterns and their corresponding alarm points for sensor 9 (outlet pressure of the fluid bed dryer)

VI. Discussion

The LSTM shows to be the most robust forecasting model. There was a clear increase in error when taking a univariate approach, highlighting the benefit of including multiple sensors from various related unit operations across the entire process for forecasting. With regards to short-term forecasting, there was negligible differences between using related, matrix and all features, indicating that the LSTM can determine feature importance, which reduces the time and effort required for pre-processing and feature engineering. In comparison, the VAR and XGB models were extremely sensitive to this, unable to sufficiently minimize the weightings (the amount of causation) of extraneous features. Notably, the XGB model did not produce any useful predictions, which suggests that it is unsuitable for these forecasting scenarios, or that it has not been properly calibrated.

An LSTM is nonetheless computationally expensive to train, and clearly the training time can be greatly reduced if related features are correctly identified. This is important when deploying such a model into industry, where there is vastly more data, and the difference in time and resources required on this scale would be economically substantial. It is therefore still important to employ feature engineering strategies for optimization and efficiency, even if an LSTM does not necessarily require it, also highlighting the importance of having domain knowledge of the system that is being modelled.

The LSTM was also able forecast long-term degradation, accurately predicting when different alarms would trigger, extrapolating varying degradation patterns up to 90 days in advance from only 18 days' worth of training data, forecasting alarms triggering with an average error of 2 days. This is in comparison to simpler ML (VAR) and linear models, which had larger errors (15 and 20 days respectively) and was incapable at deriving more complex patterns such as increasing instability.

The output forecast of the LSTM was also reliably classified, giving an average F1 score of 0.757. Again, the linear and VAR models were unable to provide the sufficient details in their respective forecasts to enable them to also be classified. The F1 score of sigmoidal degradation was, however, particularly low, despite the LSTM output clearly matching the real data. This is doubtlessly due to the classification model itself, and a better classification model is needed. Regardless, by providing a precise forecast, the LSTM provides a framework for PdM techniques, as the time until fault can be estimated, as well as the exact pattern leading to this fault, and so future faults can be individually characterized based on their patterns, and appropriate recommendations can be made to mediate this.

VII. Limitations & Future Work

Validation Against Real Industrial Data

Clearly, the biggest drawback in this approach is that the data for testing longer term degradation was simulated. While efforts have been made to replicate industry standard data as much as possible, inherently, this data will never be perfect. Nonetheless, the accuracy of the models used here would likely, increase using a real dataset. This is because degradation was simulated only using a single sensor; a process unit degrading in a real scenario is accompanied by the degradation of other unit operations in the process too, which has not been considered. This would provide further information to the models and improve performance. Machine learning models can only be as good as the data they are trained on and validating these models against real-world data will be essential in the future.

Developing "Causation" relationships

Finding and characterizing the "causation" is using the derived relationships between each sensor in a model and using it in such as manner like recommending the best approach to recover during abnormal operation. In the case of the LSTM, each feature will have been assigned a weight to determine its impact on the forecasted feature. If current degradation or atypical operation is primarily driven by an inlet temperature and other features (identification of root causes), by forecasting the effects of adjusting these, this degradation could be slowed down, to provide more time to schedule maintenance, or even prevent the incoming failure entirely. Many faults can be attributed to step changes in operational variables and forecasting the impact and determining the "causation" would prove extremely valuable.

Further Process Optimization

Currently these forecasting models provide insight into when a unit will fail, allowing for maintenance scheduling to be optimized such that less time is wasted on unplanned downtime and faults can be better anticipated, but this can be extended by optimizing based on the cost impact of this degradation. For example, a pump loses efficiency over time, and hence becomes increasingly more expensive to operate, but this cost impact of increasing degradation can be weighed against the cost of maintenance with the objective to minimize a cost function, and schedule maintenance based on the most cost-effective degradation point (as opposed to simple replacement or maintenance before it fails).

VIII. Conclusion

This paper aimed to develop methods to leverage the vast amounts of available industrial data and realize value from it through building effective machine learning models. With this research, a basis has been established for more advanced PdM techniques, comparing different ML algorithms, with the best of these models (LSTM) capable of predicting intricate patterns of degradation on DiPP data 70 days in advance. This provides more accurate forecasting, and a more precise time until fault prediction, reducing the amount of unnecessary maintenance, and any associated costs. Moreover, the model is sufficiently complex to be able to forecast non-linear patterns, volatility, and noise. On average, the LSTM could predict an emergency-level alarm tripping within 2 days of it's true time, while the VAR and Linear models took 15 and 20 days respectively, with the XGB model proven to be ineffective for long-term forecasting. Additionally, the LSTM forecasts are detailed to such an extent that they can be classified accurately, with an average F1 score of 0.76 using a sequential LSTM classifier, providing a strong framework for new prescriptive solutions. Industry standard data has been replicated to the extent that it is possible, but it is imperative these models are validated and tested against real industrial data to ensure their effectiveness. Continuing this work into the future, this research could lead to new and novel PdM solutions and accelerate the speed of industrial progress towards a smarter future, that is, Industry 4.0

References

[1] L. Wood, "Global Predictive Maintenance Market Opportunity Analysis and Industry Forecasts," 2022. [Online]. Available: https://www.businesswire.com/news/home/20211022005161

[2] Yeardley, Aaron & Ejeh et al. (2022). Integrating machine learning techniques into optimal maintenance scheduling. *Computers & Chemical Engineering*. 166. 107958. 10.1016/j.compchemeng.2022.107958.

[3] M. Abdel-Hameed, "Degradation Processes" *Advances in Degradation Modeling,* 2010.

[4] T. P. Carvalho, F. Soares and R. Vita, "A systematic literature review of machine learning methods applied to predictive maintenance," *Computers & Industrial Engineering,* 2019.

[5] S. Orhan and N. Akturk, "Vibration monitoring for defect diagnosis of rolling element bearings as a predictive maintenance tool: Comprehensive case studies," *NDT & E International,* pp. 293-298.

[6] M. Zandi, 2019, Preparing students for 21st Century Chemical Industry, Perspectives from academia and industry for dealing with a digital world, University College London, UK

91f Hallucinating Inexpensive, Diverse and Native-like Antibodies with Deep Learning

Sai Pooja Mahajan

Antibodies recognize and bind an extremely diverse repertoire of antigens via 6 hypervariable loop regions known as the complementarity determining regions or the CDRs. A common goal in antibody engineering to generate a library of CDRs enriched in binders also known as affinity maturation. Experimental methods for affinity maturation are expensive, laborious, and time-consuming and rarely allow the efficient exploration of the full design space. Conventional computational methods are relatively inexpensive but marred by inaccurate scoring functions and slow and inefficient sampling of the full combinatorial sequence space. Deep learning (DL) models are transforming the field of protein structure-prediction, engineering, and design.[1] While several DL-based protein design methods have shown promise, specialized models for antibody design are highly desirable.[2] Our approach aims to fully leverage structural information which is becoming increasingly abundant in the era of highly accurate structure prediction models such as AlphaFold[3] (all proteins, multimers) and DeepAb[4] (antibodies only). Inspired by the hallucination frameworks that specifically leverage such structure prediction DL models, we propose the F_vHallucinator for generating antibody sequences conditioned on a target antibody structure with DeepAb.[2,5,6] On a benchmark set of 60 antibodies, the F_vHallucinator recovers over 50% of the wildtype CDR sequence on all six CDR loops. At the V_H-V_L interface, the F_vHallucinator designs amino acid substitutions that are highly enriched in human repertoire sequences. Furthermore, when compared to a large experimentally characterized library of CDR H3s of the anti-HER2 antibody,[7] trastuzumab, the F_vHallucinator designs exhibit high sequence identity to known HER2-binders. We propose a pipeline that screens F_vHallucinator designs to obtain a virtually screened library enriched in binders for an antigen of interest. We apply this pipeline to the CDR H3 of the trastuzumab-HER2 complex to generate designs that retain the original binding mode and improve the binding affinity and interfacial properties. Thus, the F_vHallucinator pipeline enables fast and inexpensive generation of diverse, structure-conditioned antibody libraries enriched in binders.

1. Gao, W., Mahajan, S. P., Sulam, J. andamp; Gray, J. J. Deep Learning in Protein Structural Modeling and Design. *Patterns* **1**, 100142 (2020).
2. Norn, C. *et al.* Protein sequence design by explicit energy landscape optimization. doi:10.1101/2020.07.23.218917.
3. Jumper, J. *et al.* Highly accurate protein structure prediction with AlphaFold. *Nature* (2021) doi:10.1038/s41586-021-03819-2.
4. Ruffolo, J. A., Sulam, J. andamp; Gray, J. J. Antibody structure prediction using interpretable deep learning. *Patterns* **3**, 100406 (2022).
5. Anishchenko, I. *et al.* De novo protein design by deep network hallucination. *Nature* 1–19 (2020).
6. Wang, J. *et al.* Deep learning methods for designing proteins scaffolding functional sites. *bioRxiv* 2021.11.10.468128 (2021).
7. Mason, D. M. *et al.* Optimization of therapeutic antibodies by predicting antigen specificity from antibody sequence via deep learning. *Nat. Biomed. Eng.* (2021) doi:10.1038/s41551-021-00699-9.

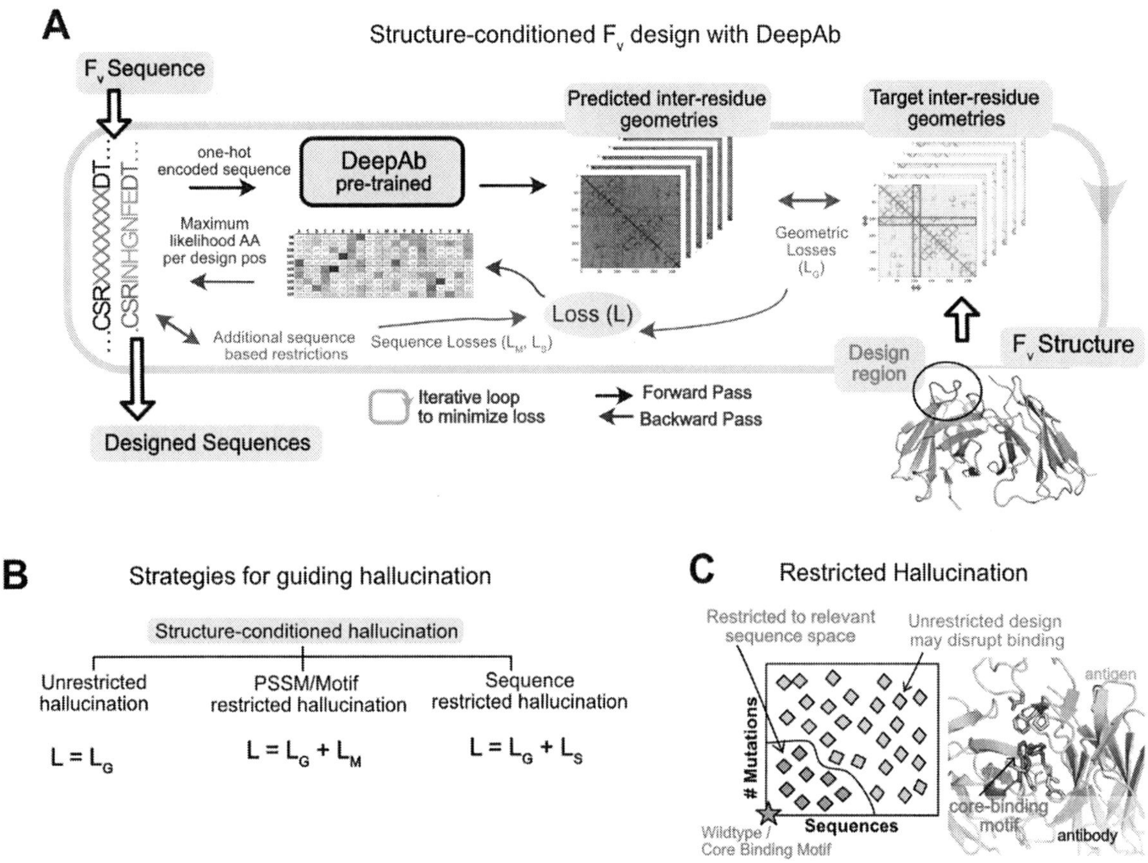

Figure 1. Framework for structure-conditioned hallucination of antibody subsequences

91g Forecasting and Reconstructing Chaotic Dynamics from Partial Observable Data

Charles Young, Michael Graham

Many applications require the prediction of a time series from only observable data. Experimental measurements of time series are often scalar or poorly spatially resolved, causing challenges in reconstructing and forecasting the system from only the current state. Takens theorem proves that for data lying on a smooth compact manifold the full state can be reconstructed from an embedding of time delayed state observations. These delay coordinates embed the attractor up to a diffeomorphic transformation. While powerful in principle, it is often challenging to learn these functions in practice, particularly for chaotic and highly nonlinear systems. We use artificial neural networks (ANNs) to learn discrete and continuous time forecasting functions as well as reconstruction functions from delay coordinate embeddings of partial observable data. We test our approach on the Lorenz system and the Kuramoto-Sivashinsky equation (KSE). We find the Lorenz attractor can be forecasted and reconstructed from a scalar observable with 2-3 time delays. For the more chaotic KSE, the attractor can be reconstructed from multivariate observations with dimensionality as low as one half the manifold dimension. Our approach provides good short time tracking and longtime statistics with relatively few embedding parameters and ANN hyperparameters.

528a Process Design of a Fully Integrated Continuous Biopharmaceutical Process Using Economic and Ecological Impact Assessment

Chaoying Ding, Hiren D. Ardeshna, Christopher Gillespie, Marianthi Ierapetritou

The majority of commercial-scale production of biological products is currently carried out in batch operation, that leads to some significant inefficiencies such as low productivity, low yield, batch-to-batch variation in product quality, and high capital and operating cost. With the rapid expansion of market demand, competition from biosimilars, and market pressure for lower drug prices, pharmaceutical industry is investigating different production alternatives such as continuous manufacturing [1-3]. Continuous biomanufacturing is a more cost-effective and flexible platform with higher productivity and yield, more consistent and better product quality, and smaller footprints [4, 5]. Although there is no fully integrated continuous platform, there have been some major advances recently [6, 7]. For the upstream process, intensified seed expansion and continuously operated perfusion bioreactors with different cell retention devices have been introduced [8, 9]. In terms of the downstream operations, multi-column periodic counter-current chromatography (PCC) for primary capture and polishing steps, different types of reactors (like coiled flow inversion reactor and plug flow reactor) for the continuous virus inactivation, and single-pass tangential flow filtration (SPTFF) for the final formulation have been developed [10-12]. There are also several successful lab-scale integrations of continuous perfusion cell culture with continuous downstream operations [13, 14].

To facilitate the commercialization of continuous manufacturing, demonstration of economic and environmental viability is important. Many research groups have shown that there are tremendous economic benefits when the operation is transitioned from batch to continuous [15-18]. Continuous platforms with different unit procedures, including perfusion techniques, capture scenarios, and membrane chromatography, have been systematically assessed [19, 20]. The ecological and operational robustness benefits of continuous biomanufacturing have been illustrated [21]. However, to the best of our knowledge, few papers have considered the media and buffer preparation when constructing continuous platforms, although studies have shown that buffer preparation can be significant in the overall operational and scheduling activities [22, 23]. The study of the bottleneck shifting with respect to the increase of upstream production is also limited. Furthermore, most papers only examine the effects of process variables on the economics when conducting scenario analysis [24, 25], but few investigates the impacts on the environmental footprints [20, 26].

This work aims to establish an in-silico fully integrated continuous platform for the production of monoclonal antibody (mAb). This developed platform incorporates media and buffer preparation steps and combine innovative technologies such as intensified seed expansion and continuous high cell density perfusion bioreactors, single-pass tangential flow filtration, and single-use technologies. After process design and development, scheduling is performed to achieve the continuous operation and real-time media and buffer addition, followed by economic and environmental analysis based on environmental indicators (E-factors). Scenario analysis is carried out to assess the effect of upstream titer and bioreactor scale on process economics and environment. The shift of process bottlenecks with the increase of the upstream production capacity is explored and analyzed. As an emerging technology, membrane chromatography can help eliminate pore diffusion, enhance the bio-separation efficiency, and lower buffer consumption [27], so membrane chromatography is incorporated into the process design and its impact on the overall process performance is investigated in terms of economic and ecological benefits. The results will provide insights into the feasibility of transitioning from batch to continuous biomanufacturing from cost and environmental aspects.

References

1. Zydney, A. L., Perspectives on integrated continuous bioprocessing—opportunities and challenges. *Curr. Opin. Chem. Eng.* **2015,** *10*, 8-13.
2. Erickson, J.; Baker, J.; Barrett, S.; Brady, C.; Brower, M.; Carbonell, R.; Charlebois, T.; Coffman, J.; Connell-Crowley, L.; Coolbaugh, M.; Fallon, E.; Garr, E.; Gillespie, C.; Hart, R.; Haug, A.; Nyberg, G.; Phillips, M.; Pollard, D.; Qadan, M.; Ramos, I.; Rogers, K.; Schaefer, G.; Walther, J.; Lee, K., End-to-end collaboration to transform biopharmaceutical development and manufacturing. *Biotechnol Bioeng* **2021**.
3. Ding, C.; Ierapetritou, M., A novel framework of surrogate-based feasibility analysis for establishing design space of twin-column continuous chromatography. *International Journal of Pharmaceutics* **2021,** *609*, 121161.
4. Yang, O.; Qadan, M.; Ierapetritou, M., Economic Analysis of Batch and Continuous Biopharmaceutical Antibody Production: A Review. *J. Pharm. Innov.* **2019,** *14*, 1-19.
5. Croughan, M. S.; Konstantinov, K. B.; Cooney, C., The future of industrial bioprocessing: batch or continuous? *Biotechnol. Bioeng.* **2015,** *112* (4), 648-51.
6. Chen, Y.; Yang, O.; Sampat, C.; Bhalode, P.; Ramachandran, R.; Ierapetritou, M., Digital Twins in Pharmaceutical and Biopharmaceutical Manufacturing: A Literature Review. *Processes* **2020,** *8* (9).
7. Khanal, O.; Lenhoff, A. M., Developments and opportunities in continuous biopharmaceutical manufacturing. *MAbs* **2021,** *13* (1), 1903664.
8. Bielser, J. M.; Wolf, M.; Souquet, J.; Broly, H.; Morbidelli, M., Perfusion mammalian cell culture for recombinant protein manufacturing - A critical review. *Biotechnol. Adv.* **2018,** *36* (4), 1328-1340.
9. Walther, J.; Lu, J.; Hollenbach, M.; Yu, M.; Hwang, C.; McLarty, J.; Brower, K., Perfusion cell culture decreases process and product heterogeneity in a head-to-head comparison with fed-batch. *Biotechnology journal* **2019,** *14* (2), 1700733.
10. Gomis-Fons, J.; Andersson, N.; Nilsson, B., Optimization study on periodic counter-current chromatography integrated in a monoclonal antibody downstream process. *J. Chromatogr. A* **2020,** *1621*, 461055.
11. Gillespie, C.; Holstein, M.; Mullin, L.; Cotoni, K.; Tuccelli, R.; Caulmare, J.; Greenhalgh, P., Continuous In-Line Virus Inactivation for Next Generation Bioprocessing. *Biotechnol J* **2019,** *14* (2), e1700718.
12. Jabra, M. G.; Yehl, C. J.; Zydney, A. L., Multistage continuous countercurrent diafiltration for formulation of monoclonal antibodies. *Biotechnol Prog* **2019,** *35* (4), e2810.
13. Karst, D. J., Steinebach, F., Soos, M., Morbidelli, M. , Process performance and product quality in an integrated continuous antibody production process. *Biotechnol. Bioeng.* **2017,** *114* (2), 298-307.
14. Coffman, J.; Bibbo, K.; Brower, M.; Forbes, R.; Guros, N.; Horowski, B.; Lu, R.; Mahajan, R.; Patil, U.; Rose, S.; Shultz, J., The Design Basis for the Integrated and Continuous Biomanufacturing Framework. *Biotechnol Bioeng* **2021**.
15. Xenopoulos, A., A new, integrated, continuous purification process template for monoclonal antibodies: Process modeling and cost of goods studies. *J Biotechnol* **2015,** *213*, 42-53.
16. Arnold, L.; Lee, K.; Rucker-Pezzini, J.; Lee, J. H., Implementation of Fully Integrated Continuous Antibody Processing: Effects on Productivity and COGm. *Biotechnol. J.* **2019,** *14* (2), e1800061.
17. Grilo, A. L.; Mateus, M.; Aires-Barros, M. R.; Azevedo, A. M., Monoclonal Antibodies

Production Platforms: An Opportunity Study of a Non-Protein-A Chromatographic Platform Based on Process Economics. *Biotechnol J* **2017,** *12* (12).
18. Nandi, S.; Kwong, A. T.; Holtz, B. R.; Erwin, R. L.; Marcel, S.; McDonald, K. A., Techno-economic analysis of a transient plant-based platform for monoclonal antibody production. *MAbs* **2016,** *8* (8), 1456-1466.
19. Pollock, J.; Ho, S. V.; Farid, S. S., Fed-batch and perfusion culture processes: economic, environmental, and operational feasibility under uncertainty. *Biotechnol. Bioeng.* **2013,** *110* (1), 206-19.
20. Cataldo, A. L.; Burgstaller, D.; Hribar, G.; Jungbauer, A.; Satzer, P., Economics and ecology: Modelling of continuous primary recovery and capture scenarios for recombinant antibody production. *J Biotechnol* **2020,** *308*, 87-95.
21. Pollock, J.; Coffman, J.; Ho, S. V.; Farid, S. S., Integrated continuous bioprocessing: Economic, operational, and environmental feasibility for clinical and commercial antibody manufacture. *Biotechnol Prog* **2017,** *33* (4), 854-866.
22. Petrides, D.; Carmichael, D.; Siletti, C.; Koulouris, A., Biopharmaceutical Process Optimization with Simulation and Scheduling Tools. *Bioengineering (Basel)* **2014,** *1* (4), 154-187.
23. Hummel, J.; Pagkaliwangan, M.; Gjoka, X.; Davidovits, T.; Stock, R.; Ransohoff, T.; Gantier, R.; Schofield, M., Modeling the downstream processing of monoclonal antibodies reveals cost advantages for continuous methods for a broad range of manufacturing scales. *Biotechnology journal* **2019,** *14* (2), 1700665.
24. Yang, O.; Prabhu, S.; Ierapetritou, M., Comparison between Batch and Continuous Monoclonal Antibody Production and Economic Analysis. *Ind. Eng. Chem. Res.* **2019,** *58* (15), 5851-5863.
25. Gupta, P.; Kateja, N.; Mishra, S.; Kaur, H.; Rathore, A. S., Economic assessment of continuous processing for manufacturing of biotherapeutics. *Biotechnol Prog* **2021,** *37* (2), e3108.
26. Madabhushi, S. R.; Gavin, J.; Xu, S.; Cutler, C.; Chmielowski, R.; Rayfield, W.; Tugcu, N.; Chen, H., Quantitative assessment of environmental impact of biologics manufacturing using process mass intensity analysis. *Biotechnol Prog* **2018,** *34* (6), 1566-1573.
27. Orr, V.; Zhong, L.; Moo-Young, M.; Chou, C. P., Recent advances in bioprocessing application of membrane chromatography. *Biotechnol Adv* **2013,** *31* (4), 450-65.

528b An Integrated State Estimation, Covariance Estimation, and Optimal Control Framework of a Semi-Batch Reactor for Bioprocess Applications

Ronald Alexander, Marcelo P. A. Ribeiro, Fernando V. Lima

Despite improvements in process model development regarding high-fidelity first-principles models and machine learning-based advances, it is still extremely challenging to accurately model real chemical and biochemical processes. Unmodeled disturbances such as fouling, environmental heat losses, pH variations, or even poor storage of enzymes can influence reaction rates, chemical equilibrium, and other critical process variables. Ultimately, such unmodelled and unaccounted disturbances can cause significant departure between the available plant model and the true process states, which makes it difficult to solely rely on the developed process models for process systems engineering (PSE) applications. Although real-time and delayed process measurements are available to track an ongoing process, these measurements are often corrupted by significant amounts of noise, poor calibration, or are unable to directly quantify key process states such as concentrations [1]. Traditionally, state estimation approaches such as the Extended Kalman Filter (EKF) or Moving Horizon Estimation (MHE) have been implemented to blend these process models and measurements together to derive more robust approximations of the true process states. However, the performance of these estimation techniques is highly dependent on the correct selection of the noise covariance matrices, which is a challenging endeavor that is typically solved via trial-and-error procedures in real process applications. These challenges hinder the implementation of other PSE methods such as advanced control as reliable insight into the true process states is required for optimal trajectory tracking and model predictive control (MPC).

In this work, an integrated state estimation, covariance estimation, and optimal control framework is proposed. Such framework provides robust state estimates that can be used by the optimal control algorithm to improve process performance and remediate noise effects. Due to the complex estimation challenges and increasing market demand, this framework is applied to a batch lactose to galacto-oligosaccharide (GOS) system utilizing β-galactosidase [2]. GOS typically refers to a class of lactose derived products ranging in the size of 2-5 units of galactose sometimes attached to a glucose residue. These products have important prebiotic and health positive properties and can be used in many food products such as infant formula [3]. In this framework, state estimation is handled using various EKF and MHE formulations that are selected based upon the specific needs of the process. The noise covariance matrices are defined using the autocovariance least-squares (ALS) technique that has previously exhibited excellent performance on both continuous and batch process applications [4]. These two techniques provide accurate values of the process states and allow the optimal control algorithm implementation. By combining these three parts into a single framework, it is possible to overcome plant-model mismatch, measurement and process noises.

To challenge this framework, various sources of noises are introduced into the GOS system by varying kinetic parameter values, changing feedstock concentrations, and applying white noises to the process measurements. Overall, these sources of noise provide a simulation of real process disturbances and allow this framework to be applied to a more realistic system. Furthermore, due to the flexibility of this noise generation approach and the combined framework proposed, this ultimately provides a customizable solution to addressing process monitoring and control challenges faced by the modern chemical and biochemical industry.

References

[1] Alexander, R., Campani, G., Dinh, S., Lima, F.V. (2020). Challenges and Opportunities on Nonlinear State Estimation of Chemical and Biochemical Processes. *Processes, **8**, 1462.*

[2] Schultz, G., Alexander, R., Lima, F. V., Giordano, R. C., Ribeiro, M. P. A. (2021). Kinetic Modeling of the Enzymatic Synthesis of Galacto-Oligosaccharides: Describing Galactobiose Formation. *Food and Bioproducts Processing, **127***, 1-13.

[3] Lamsal, B. (2012). Production, health aspects and potential food uses of dairy prebiotic galactooligosaccharides. *J Sci Food Agric, **92***, 2020-2028.

[4] Rinćon, F., Roux, G.A., Lima, F. V. (2014). The Autocovariance Least-Squares Method for Batch Processes: Application to Experimental Chemical Systems. *Ind. Eng. Chem. Res., **2014***, 18005-18015.

528c Hybrid Kinetic-Stoichiometric Model of CHO Cell Fed-Batch Process

Mariana Monteiro, Cleo Kontoravdi

Mammalian cells are used to produce half of the commercially available therapeutic proteins, with Chinese Hamster Ovary (CHO) cells representing the major fraction of the used cells. CHO cell systems enable the application of certain post-translational modifications to obtain specific structures of protein products that are compatible with humans. Production processes involving CHO cell lines typically operate in fed-batch mode, as it allows cells to be fed throughout the cycle, thus preventing the culture from nutrient depletion before production goals are met.

Process models of fed-batch CHO cell cultures offer little predictive power due to having been fitted for a particular set of culture and process conditions. Whenever process or culture conditions change, the models must be updated with new experimental data, and new kinetic parameters must be estimated. This works aims to reduce this burden by developing a hybrid kinetic-stoichiometric model, based on a state-of-the-art genome-scale metabolic model.

A single CHO cell contains thousands of metabolites, genes and reactions. Thus, including a full description of the metabolism of a CHO cell would make the process model computationally intractable. As a result, this work also investigates the trade-off between the granularity, the computational burden and the predictive accuracy of metabolic network models.

The proposed hybrid model combines a fed-batch CHO cell culture model with a CHO cell metabolic model. The model of the fed-batch process is given by the total biomass and key metabolites balance, written as a system of ordinary differential equations. The metabolic model takes the form of a genome-scale model, which is a constraint-based model, that relies on a system of linear equations to describe the mass balance of each metabolite. Given the high degrees of freedom, the system of equations is solved as an optimization problem, where the metabolic fluxes are constrained by experimental data.

The proposed framework establishes the link between the metabolic and process models via the fluxes of secreted metabolites. The dynamic formulation of the metabolic model provides the fluxes of secreted metabolites, biomass and target protein produced as a function of time. By having a set of shared variables, we are able to connect both the fed-batch and metabolic models.

The proposed framework is capable of simulating a fed-batch CHO cell culture process, with an embedded dynamic genome-scale model. Given a starting state of process conditions, the model is also able to predict the uptake and secretion rates of key metabolites, as well as the biomass and protein produced.

In this work, we propose a hybrid modelling framework that overcomes the parameter estimation burden of fitting several kinetic parameters, by taking advantage of state-of-the-art genome-scale models. This represents a step forward for accurately modelling fed-batch CHO cell cultures.

528d Efficient Modeling of Critical Process Parameters in Bioreactors – a Case-Study across Scales

Johannes G. Khinast, Philipp Eibl, Christian Witz

Critical process parameters during operation of industrial biofermenters such as volumetric mass transfer coefficient, gas-holdup, dissolved oxygen or sufficient supply of nutrients, are highly dependent on the fermenter geometry and scale and cannot be easily predicted by empirical correlations alone, which is apparent by the vast number of existing correlations and variations of up to orders of magnitude [1]. Ensuring optimal environmental conditions for microorganisms in large-scale production fermenters in the biopharmaceutical industry is of vital importance for the safety, stability, and optimized productivity of the fermentation process.

In this work, we present an in-silico approach to predict these critical parameters based on first principles using the commercial software SIMVANTAGE (simvantage.com), that utilizes the Lattice Boltzmann Method (LBM) for transient simulations of the liquid phase. It is based on an in-silico scale-independent simulation method to predict key parameters based on first principles, even suitable for large-scale fermenters. The implementation is optimized for state-of-the-art Graphics Processing Units (GPUs) and includes the motion, breakup, and coalescence of the dispersed gas phase via the Euler-Lagrange method, transport phenomena between phases and throughout the fermentation broth, as well as the movement and distribution of microorganisms. The inclusion of the microorganism movement allows for a detailed analysis of the inhomogeneities within the fermentation broth by tracking the individual organism lifelines and the environmental conditions of the whole population over time [2].

The first part of the presentation focusses on the characterization of a lab-scale reactor to show the applicability of the solver for capturing a multitude of operational regimes from the vortex Cavity regime to the loaded and flooded regime. In the second stage, results are compared to geometrically similar large-scale fermenters to assess the scale-independency of the simulation framework without re-tuning of modeling parameters.

Then, a use-case is shown on how to transfer process conditions from the lab scale to the production scale based on the similarity of the environmental conditions of the organism population. Possible applications include reducing experimental load during scale-up, as well as the construction of representative scale-down models of large-scale fermenters.

[1] Yawalkar et al. "Gas-Liquid Mass Transfer Coefficient in Stirred Tank Reactos", Canadian Journal of Chemical Engineering, Volume 80, 2002, 840-848

[2] Haringa et al. "Euler-Lagrange computational fluid dynamics for (bio)reactor scale down: An analysis of organism lifelines" Engineering in Life Sciences, Volume 16 (7), 2016, 652-663

Title: Tablet Manufacturing Using Dried Whole Puerto Rican Sweet Potato and Breadfruit Particles

Authors: **Luis F. Torrens Sotomayor** *(Presenting Author)*
University of Puerto Rico-Mayagüez Campus - Department of Chemical Engineering. Center for Pharmaceutical Engineering Development and Learning
luis.torrens@upr.edu

Carlos Velázquez Figueroa
University of Puerto Rico-Mayagüez Campus - Department of Chemical Engineering. Center for Pharmaceutical Engineering Development and Learning
carlos.velazquez9@upr.edu

Abstract Text:

In Puerto Rico, there is a serious need to develop alternatives for effective recovery after a catastrophic event, including the strong dependance of importation of products and storage facilities. One industry that has constantly being affected has been the pharmaceutical industry since most of its raw ingredients for manufacture comes from foreign industries. An alternative is to encourage the production of these ingredients using local resources as a source. An area of opportunity withing these ingredients is the development of novel ingredients. In fact, current FDA approaches are aimed at promoting new formulations using novel excipients that had bast experience in other industries, have a positive functional effect on the patient, and can promote the creation of new drugs.

Local crops harvested locally can be a source for these novel ingredients, which although historically have been widely used for starch extraction, there is a lack of information on using the whole food components for drug formulation. This research focused on the use of dried whole-food particles obtained from local sweet potato and breadfruit in the preparation of food-based tablets. For this, the effect of varying the type of raw material (sweet potato and breadfruit), the method of cutting the raw material (cubes and chips), the addition of a binder to the formulation (0% and 2% PVP), and different compression forces (4, 6, and 8 metric tons) were studied to understand the effect on the tablet overall performance using common characterizations methods for tablet manufacture. To understand the correlation between the variables and the results, analysis between results were done using Fisher analysis. General results showed that the compression force and agglomerate percentage had is significant effect on the tablets friability and hardness while the material composition, the method of cutting, the compression force, and the agglomerate percentage all had a significant effect on the disintegration time. For the friability test, only the sweet potato tablets with 2 % of PVP reached values below the 1% of weight loss. For the tablet hardness, most of the tablets had hardness greater than 4kP except for the case tablets compressed at 4MT with 0% PVP and whose raw materials were cut into chips. Mean disintegration times for sweet potato ranged from 7 to 18 minutes while breadfruit tablets took 3 to 5 minutes to disintegrate. Dissolution profiles had the expected behavior of a constant increasement in concentration until it reached concentration equilibrium before 15 minutes. These results shows that it is feasible to compact these whole foods particles and can be later studied for future formulations. Next steps are to improve the particle properties with minimal physical modifications and study the mixture between different foods. Now, the findings of this research can promote more investigation these particles and provide a product which can, in addition to being used in other applications in the food industry for example, can be used as a novel excipient that can reduce import and transportation costs, minimize storage dependency, and even as an emergency food source.

528f Explainable One-Class Classification Neural Network Model for Tablet Quality Control

Alex Taylor, Rob Holt, Aycan Hacioglu

Typical quality control tests applied to tablets include the content of the active ingredient or absolute drug content tests, uniformity of weight, uniformity of the content, disintegration time test, and dissolution tests. Using visual inspection systems, tablets are also inspected for cracks, chips, unusual colors, and other defects.

AI-driven methods are increasingly used in manufacturing processes for quality control as they promise less manual work and higher accuracies than human assessment. In this talk, we present an application of an explainable one-class neural network model for tablet quality control. With a one-class learning approach, accurate models could be developed even if the anomalies are scarce or change over time. Additionally, with this approach, smaller amounts of data are needed, and the labeling effort required is low – the knowledge of "good" and "bad" images is sufficient. The model is trained in MATLAB with images of normal pills and synthetically generated anomaly images. Following the recommended Fully Convolutional Data Description (FCDD) approach, a couple of realistic examples of anomaly images were added to the training data as outlier exposure to improve learning. The model produces a heatmap with the probability that each pixel is anomalous. The classifier labels images as normal or anomalous based on the mean value of the anomaly score heatmap. The model yields 97% accuracy in its classifications. We also demonstrate how similar AI models can be deployed to larger production systems.

528g Closed-Loop Optimal Control Operation of an Industry-Scale Bioreactor and Experimental Validation

Parth Shah, M.Ziyan Sheriff, Mohammed Saad Faizan Bangi, Joseph Kwon, Costas Kravaris, Chiranjivi Botre, Junichi Hirota

Chemical processes worldwide operate in a way that maximizes the profitability while minimizing the material costs. Currently, most industrial plants achieve their targets by manipulating the inputs heuristically. But in recent times, there has been a transition to industry 4.0 where tools like artificial intelligence, machine learning, and real-time optimal control are utilized to achieve efficient and improved performance [1]. As a case study, we have developed an online controller that determines the optimal operating conditions for an industrial-scale bioreactor while considering process constraints. Controlling a fermenter is challenging due to the high sensitivity of the micro-organisms to slight changes in inputs and the culture media [2]. Hence, a sophisticated model predictive control (MPC) scheme is needed that optimizes the operation of the fermenter, reaching the product and cost targets while also ensuring the safety of the process by considering plant constraints [3].

Optimal control of any process is highly dependent on its accurate modeling. For this case study, we utilize a recently developed hybrid model for a biochemical fermenter [4]. This model reasonably predicts the process states like biomass and substrates in both batch and fed-batch operations. A multi-objective control problem is solved, which maximizes the product amount, minimizes the operating cost, and maintains the substrate at a fixed concentration. The optimization problem is set up in GAMS, and a framework is developed that links it to the graphical user interface (GUI), where the operator can feed real-time values [5]. This shrinking horizon MPC is re-initialized with the actual plant values of the internal states and inputs (during abnormal operation) through this framework. The developed algorithm effectively determined the optimal input trajectories of temperature and flow rates while ensuring that the bounds for the inputs and their rate of change are maintained. This algorithm uses the hybrid model for the state predictions for the corresponding manipulated inputs and was shown to be within the normal operating range. This approach was then validated using multiple batches of experimental data from an industry-scale bioreactor, and an overall 8% increase in productivity was achieved. The closed-loop operation of this framework helps in re-initializing the targets at every instant, and the MPC runs after every sampling time, providing the operator with the next set of inputs to be applied to the bioreactor.

The developed MPC algorithm is computationally inexpensive, accounts for batch-to-batch variability, and can improve the profitability of the process by determining the optimal operating conditions. A framework is also developed that links the operator to the machine using a GUI, helping in real-time monitoring of the process. The control algorithm is implemented on an industrial-scale bioreactor, and the performance of the proposed framework is demonstrated.

References:

1. Sansana, J., Joswiak, M. N., Castillo, I., Wang, Z., Rendall, R., Chiang, L. H., andamp; Reis, M. S. (2021). Recent trends on hybrid modeling for Industry 4.0. Computers andamp; Chemical Engineering, 151, 107365.
2. Petersen, L. N., andamp; Jørgensen, J. B. (2014, June). Real-time economic optimization for a fermentation process using model predictive control. In 2014 European Control Conference (ECC) (pp. 1831-1836). IEEE.
3. Nagy, Z. K., andamp; Braatz, R. D. (2003). Robust nonlinear model predictive control of batch

processes. AIChE Journal, 49(7), 1776-1786.
4. Shah, P., Sheriff, M. Z., Bangi, M. S. F., Kravaris, C., Kwon, J. S. I., Botre, C., andamp; Hirota, J. (2022). Deep neural network-based hybrid modeling and experimental validation for an industry-scale fermentation process: Identification of time-varying dependencies among parameters. Chemical Engineering Journal, 135643.
5. Ferris, M. C., Jain, R., andamp; Dirkse, S. (2011). Gdxmrw: Interfacing gams and matlab. Online: http://www. gams. com/dd/docs/tools/gdxmrw. pdf.

420a Deep Model-Based Reinforcement Learning for Active Flow Control of Turbulent Couette Flow

Kevin Zeng, Alec Linot, Michael Graham

Design of active control strategies for turbulent drag reduction is a challenging task due to the complex nonlinear dynamics and the difficulty in devising good control targets. Deep reinforcement learning (RL), an emergent machine learning method capable of learning complex control strategies for high-dimensional systems from data, is able to address broad macroscopic control objectives, such as drag minimization, making it promising for discovering flow control strategies. However, the iterative RL process requires vast amounts of data to be generated from direct interactions with the target system. For high-dimensional and computationally demanding simulations (such as direct numerical simulations (DNS)) or fluid flow experiments this quickly becomes prohibitively expensive. We mitigate this challenge in a completely data-driven fashion by combining data-driven reduced-order models (ROM) of the flow system with deep RL. We demonstrate our method on a DNS of turbulent Couette flow modified with four independently controlled wall-normal gaussian jets on each wall for control. We set the control objective to be the minimization of the system's drag in addition to the cost of control. We learn a ROM of the high-fidelity DNS' turbulent dynamics with control by combining an undercomplete autoencoder with a neural ordinary differential equation (ODE) to learn a compressed, low-dimensional model of the original system's dynamics -- which is trained directly from collected data. This ROM is substituted in place of the high-fidelity DNS during RL training to accelerate the learning process. We then deploy our ROM-based control strategy to the original high-fidelity DNS for control validation.

420b Computational Method for Optimization of Micro-Structured Materials for Passive Cooling Applications

Andrew Cochrane, Weston Ortiz, Rekha Rao

Over $372 billion is spent on electricity for cooling annually. Passive cooling, which requires no electricity, can be used to dramatically reduce energy usage for refrigeration and other cooling requirements. Engineered materials, structured at the micro-scale, are designed to emit more radiant energy than they absorb so they can passively cool the bodies they are attached to. Optimization is performed by iterative simulation of harmonic Maxwell's equations implemented in a research finite-element-method (FEM) code. Particular analyses of several structures compare results from this implementation to other simulations used by the passive cooling research community. [2019 Gao et al.]

Solar radiation impinging on the Earth's atmosphere is approximately the spectrum of a 5800 K blackbody, spanning wavelengths from 250 nm to 2.5 um. Objects at Earth's ambient temperature (roughly 300 K) radiate at longer wavelengths in the mid- to long-wavelength IR spectrum - roughly 3 to 30 um wavelength range. Several metamaterial structures have been proposed that radiate more energy than the blackbody level by preferentially radiating in the 8-13 um window. These are nano-layered films or glass bead composite films or structured surfaces utilizing resonance from localized surface plasmon modes – often referred to as Frohlich resonance - in the 8-13 um atmospheric window. If thermal emission can be increased, energy saved from cooling films operating 24 hours/day could rival or exceed that produced by solar PV panels of the same area.

Kirchoff's law of thermal radiation states that at steady-state and for a given wavelength, a material's surface emissivity and absorptivity are the same. This law enables study of emissivity because absorbence in a thin and structured film can be modeled simply through the extinction coefficient attributed to it's bulk material. Electromagnetic and energy equations are coupled so planewave-light absorbed in the material is converted to thermal energy, resulting in an increased temperature of the structure driving heat flow through the panel. Analysis of the energy balance results in wavelength-specific values of absorptivity.

Results of a material property study of photonic crystal-structured UV-curable adhesive (PCSUVA) for passive cooling provides dielectric properties and structure geometry to create simulations of thin film and truncated cone photonic structures. Simulation results are compared to those obtained using commercial software. Sensitivity analysis of absorptivity to material properties and structure geometry shows how the metamaterial may be tuned to improve it's efficacy in the application of passive cooling.

[2019 Gao et al.] Gao, M.; Han, X.; Chan, F.; Zhou, W.; Liu, P.; Shan, Y.; Chen, Y.; Li, J.; Zhang, R.; Wang, S.; Zhang, Q.; Zheng, W. Approach to fabricating high-performance cooler with near-ideal emissive spectrum for above-ambient air temperature radiative cooling. Solar Energy Materials and Solar Cells 2019, 200.

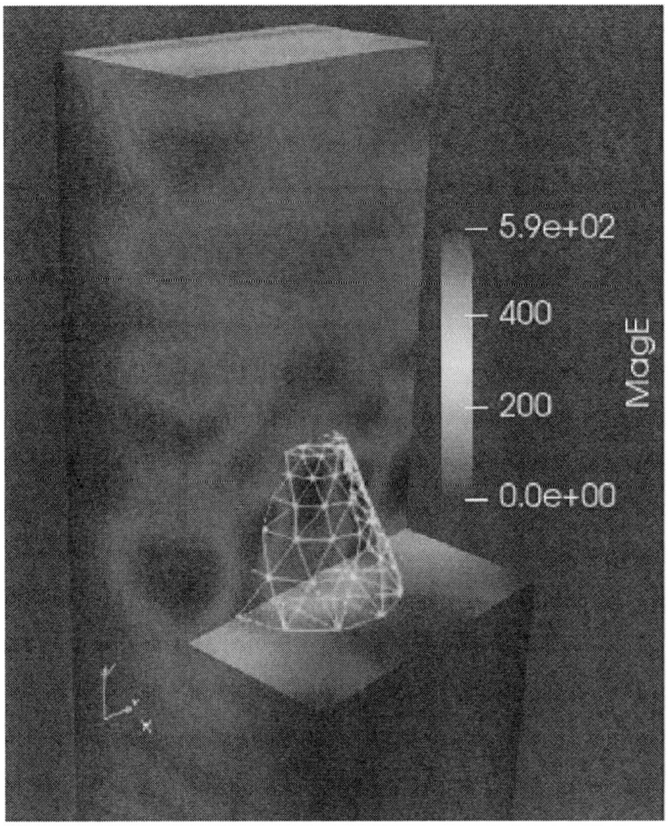

5.9e+02
400
200
0.0e+00
MagE

420c Determining Appropriate Input Perturbation for an Improved Intensified Design of Experiments Approach for System Identification of a Continuous Bioprocess

Samardeepsingh Sarna, Nikesh Patel, Brandon Corbett, Chris McCready, Prashant Mhaskar

This work addresses the problem of determining appropriate input perturbation for cell culture bioprocesses in general, and for an monoclonal antibodies industrial bioreactor in particular. The key objective is to demonstrate the feasibility of using far more perturbations than typically done in bio process identification, although significantly less than standard chemical process system identification, to yield data rich enough for the purpose of data driven modeling (and subsequently, control) and yet not having disturbed the cell culture too much.

A proprietary mechanistic model developed by Sartorius for their Cellca cell line is first introduced to serve as a test bed. Subsequently, this test bed is used to address the question of determining the frequency of input perturbation sufficient to identify a data driven dynamic model. To this end, the test bed is used to generate data at various frequencies and a linear time invariant model is identified. The predictive capability of the identified model using limited data from only one batch is used to ascertain the frequency of changes in data generation such that the changes are acceptable from a biological standpoint, and yet generate sufficiently rich data. In particular, a frequency of perturbations at once every three days is found to balance these tradeoffs for the monoclonal antibody process under consideration. The results from the manuscript are meaningful both from a specific results standpoint (as illustrated by subsequent adoption by Sartorius), but also by providing a mechanism to ascertain such information for other bioprocesses.

420d Evaluating Implementation Feasibility of Cyberattack Detection Strategies Based on Lyapunov-Based Economic Model Predictive Control

Dominic Messina, Helen Durand

Recent works from our group have explored cyberattack resilient Lyapunov-based economic model predictive control (LEMPC) frameworks in terms of detection of attacks on controller components, specifically sensors and actuators [3,4,5]. To ensure confidence in cyberattack detection and safety guarantees however, the LEMPC must be designed to meet conditions theoretically guaranteeing feasible control actions and stable operation, subject to assumptions. Though the theoretical conditions are sufficient but not necessary, they are important from a cybersecurity standpoint for guaranteeing that there are not vulnerabilities in integrated control and detection policies which an attacker might exploit. Though there are methods for designing parts of the control strategy including the Lyapunov function [6] and explicit stabilizing controller [2], developing methods for obtaining all of the parameters and functions required for guaranteeing cybersecurity for practical systems remains a challenge.

This talk will discuss various strategies employed to approximate sets of functions and parameters which meet the control theory of LEMPC as a move toward discussing the practicality of the control-theoretic cyberattack detection policies from our prior work, and what it would take to provide these greater practical utility. We first present a discussion of determining LEMPC parameters which meet theoretical guarantees for a chemical process example involving a continuous stirred tank reactor to be placed under LEMPC [1]. We present a series of inequality conditions central to LEMPC theory [1] and select several strictly increasing functions required by the theory. We then perform a 'best-case' analysis to find an upper bound on the size of the sampling period of the LEMPC that would be required, in the absence of plant/model mismatch, for the theoretical conditions of LEMPC to be satisfied. We demonstrate that the best case can be quite small, but that it also partially depends on the functions initially selected. For this reason, we then explore simultaneous development of various functions required by the theory using an optimization problem where various functions are suggested to take a certain functional form, and the optimization problem is meant to aid in estimating the coefficients. We discuss the benefits and limitations of these methods, and what they reveal about the use of LEMPC from a practical perspective. Building from this discussion, we next review several cyberattack detection strategies from [3] from the viewpoint of showcasing how the analysis described for LEMPC might be performed for obtaining the control-theoretic parameters for these control designs. We discuss how the added consideration of cybersecurity affects the 'best-case' results obtained as the initial approximation of the parameters.

[1] Heidarinejad, Mohsen, Jinfeng Liu, and Panagiotis D. Christofides. 'Economic model predictive control of nonlinear process systems using Lyapunov techniques.' *AIChE Journal* 58.3 (2012): 855-870.

[2] Lin, Yuandan, and Eduardo D. Sontag. 'A universal formula for stabilization with bounded controls.' *Systems andamp; Control Letters* 16.6 (1991): 393-397.

[3] Oyama, Henrique, and Helen Durand. 'Integrated cyberattack detection and resilient control strategies using Lyapunov-based economic model predictive control.' *AIChE Journal* 66.12 (2020): e17084.

[4] Oyama, Henrique, Keshav Kasturi Rangan, and Helen Durand. 'Handling of stealthy sensor and

actuator cyberattacks on evolving nonlinear process systems.' *Journal of Advanced Manufacturing and Processing* 3.3 (2021): e10099.

[5] Oyama, Henrique, Dominic Messina, Keshav Kasturi Rangan, and Helen Durand. 'Lyapunov-Based Economic Model Predictive Control for Detecting and Handling Actuator and Simultaneous Sensor/Actuator Cyberattacks on Process Control Systems.' *Frontiers in Chemical Engineering*: 11.

[6] Papachristodoulou, Antonis, and Stephen Prajna. 'On the construction of Lyapunov functions using the sum of squares decomposition.' *Proceedings of the 41st IEEE Conference on Decision and Control, 2002.* Vol. 3. IEEE, 2002.

420e Ontology Engineering: Support to Decision Making in Biorefining

Franjo Cecelja, Edlira Kalemi, Nikolaos Trokanas, Linsey Koo

Biorefining is a dynamic field with ever grooving number of processes operating in different environment, using different feedstocks and providing expanding range of products. Option scoping and process synthesis and process design are becoming increasingly complex [1]. While development and integration of process and feedstock models and datasets from laboratory experiments are crucial at design stage [2], process and feedstock discovery and valorisation are important at operational stage [3]. This in turn necessitates efficient decision support tool addressing stated complexity and assuring timely and accurate decisions made at increasingly shorter time span.

This paper proposes a knowledge based decision support tool for biorefining which:

- Coordinates operation of biorefining repositories of process and feedstock models and datasets, as well as repositories of existing biorefining processes and available feedstocks, all with fast and precise discovery and audited updates;
- Supports model and data integration used for for process synthesis and design [4];
- Enables screening opportunities and integration of concomitant networks based on technological capability and resource availability for targeted products [3, 5];
- Assessment of economic and environmental potential and effects [6].

The proposed decision making tool employs ontologies to embed tacit knowledge in the domain of biorefining, knowledge gained from past experience but also from the latest research and otherwise advances in biorefining [1, 2]. More specifically, a set of integrated ontologies address i) functionality and otherwise properties of biorefining models and datasets from laboratory experimentation and biorefining processes, or simply from analytical assessments, ii) nature of feedstock, i.e. variability in composition, dynamics in availability and pricing, as well as economic and environmental properties including hazardousness, iii) processing technologies in terms of processing capabilities, but also range of type of inputs, conversion rates, water and energy requirements, range of capacities, emissions as well as fixed and operational costs and environmental effects, and iv) general knowledge and past experience in process design and synthesis. While tacit knowledge associated with all four elements is embedded in the ontology structure, the explicit knowledge associated with particular process models and datasets, as well as particular process and feedstock is enquired during the ontology instantiation with actual data collected from the model, data, process and feedstock owners during the registration. The ontologies are designed using ontology web language and hence prepared to grow and to share. In the current implementation more than 400 different model and datasets have been included on the top of more than 150 different type of feedstock and over 40 different biorefining technologies, all organised in respective repositories.

Purpose designed matchmaker [5] is used to i) provide a fast and accurate discovery of elements in all repositories and enable repository audited updates, ii) enable model and data integration, and iii) formation of processing networks with assessment of economic and environmental benefits. Semantic relevance of integrated models and datasets, as well as integrated processes into a network solution defines their mutual suitability and it is calculated from distance between the two instances in the respective ontology. Explicit relevance is calculated using vector similarity algorithm for respective properties, such as quantity, availability, geographical location and environmental and economical potential etc. More intuitive and complex model and data networks, as well as processing networks are proposed by reclusively repeating matches between two participants which in turn gives an opportunity for even better economic and environmental savings and/or targeted production. Both semantic and explicit matching relevance are aggregated in into a numerical values use for match ranking and providing beter but flexible decision support to the user.

The proposed decision making platform has been implemented as a web service with performance validated verified using several real-life cases for model and data integration, as well as real-life biorefining scenarios.

1. Bussemaker, M., et al., *Ontology Modelling for Lignocellulosic Biomass: Composition and Conversion.* Computer Aided Chemical Engineering: 28th European Symposium on Computer A, 2018.
2. Koo, L., N. Trokanas, and F. Cecelja, ***A semantic framework for enabling model integration for biorefining***. Computers andamp; Chemical Engineering, 2017. **100**: p. 219-231.
3. Raafat, T., et al., *An Ontological Approach Towards Enabling Processing Technologies Participation in Industrial Symbiosis.* Computers andamp; Chemical Engineering, 2013. **59**(1): p. 33-46.
4. Cecelja, F., et al., *Integration of CAPE Models and Data for the Domain of Biorefining: InterCAPEmodel Ontology Design.* Computers and Chemical Engineering, 2019. **- accepted for publication.**
5. Trokanas, N., F. Cecelja, and T. Raafat, *Semantic Input/Output Matching for Waste Processing in Industrial Symbiosis.* Computers andamp; Chemical Engineering, 2014. **66**(1): p. 259 - 268.
6. Trokanas, N., F. Cecelja, and T. Raafat, *Semantic Approach for Pre-assessment of Environmental Indicators in Industrial Symbiosis.* Journal of Cleaner Production, 2015. **96**(1): p. 349-361.

420f Benchmarking Model Selection Criteria within an Automated Kinetic Rate Equation Discovery Framework

Miguel Ángel de Carvalho Servia, Antonio del Rio Chanona, Dongda Zhang, Klaus Hellgardt, Mimi Hii

Accurate, predictive, and interpretable mathematical models are important from a theoretical and practical point of view. Theoretically, these models allow engineers to gain fundamental understanding of physical phenomena. Practically, these models allow engineers to optimize, control and even develop novel processes. Nevertheless, the automated discovery of true kinetic rate models remains an open challenge within the chemical engineering community. This challenge, as mentioned, has many ramifications within the industrial world, ranging from sub-optimal control to difficulties within the design and upscaling of chemical processes.

Different modelling techniques have been proposed and explored in the literature: white-box modelling (i.e.: first-principles models), grey-box modelling (i.e.: hybrid models), and black-box modelling (i.e.: data-driven models). The grey-box modelling technique aims to exploit the advantages of white-box modelling, namely its predictive ability, whilst also exploiting the advantages of black-box modelling, namely its ease of construction. However, most hybrid models presented in the literature make subjective and undetermined assumptions about the chemical system investigated (e.g.: assuming kinetic formalisms) and do not include a rigorous model selection method. While these assumptions hinder the accuracy and predictability of the model proposed, the absence of a rigorous model selection method limits the capabilities of finding the underlying ground truth of the chemical system.

Furthermore, automated kinetic rate equation discovery deals with complex nonlinear optimization problems arising from model parameter estimation. Two approaches can be used to solve the parameter estimation problem within the model building of dynamic systems framework: sequential and simultaneous. The sequential approach integrates a differential model at each iteration to judge its performance. The simultaneous approach reformulates the parameter estimation problem, usually via approximations introduced by different orthogonal collocation methods. Both approaches suffer from noteworthy shortcomings. The sequential approach suffers from high computational times caused by the integration step, whilst the simultaneous approach suffers from sub-optimal solutions caused by approximations and complex reformulations.

Due to the discussed limitations, both within the model building and parameter estimation framework, we have developed a method that tackles these issues. Our method uses minimal – but important and physically-driven – prior knowledge to guide a symbolic regression algorithm to propose kinetic rate equations for a given chemical system. Using carefully analyzed model selection criteria and model-based design of experiments, we can robustly identify and choose the model that accurately describes the system's kinetics while providing limited, but highly informative data. Additionally, because of the method's setup, the integration of the differential model is altogether avoided without the need of reformulating the parameter estimation problem.

Particular attention has been given to the analysis of the model selection method, as this is the critical step to retrieve the true model's structure of a chemical system. Over the years, many model selection criteria have been proposed. These model selection approaches have been motivated by a wide range of philosophical viewpoints, such as information criteria, Bayesian, information-theoretic or decision-theoretic perspectives. Some works in the literature use a model selection criterion to differentiate between competing models. However, the choice of a particular criterion is seemingly random.

To strengthen our automated kinetic rate model approach, we benchmarked a number of model selection criteria on different case studies, whilst varying the quantity and information content of the data provided. We also assessed the level of noise that each criterion was able to withstand until it started selecting wrong models. Our objective with this study was to discover which criterion, if any, was better suited for the kinetic rate discovery task, as well as investigating which criterion was the most robust. Our study demonstrated that, from the criteria examined, the Hannan and Quinn criterion is the most robust and well-suited model selection criterion for the problem class at hand. All in all, our meticulous choice of model selection criterion integrated within our proposed methodological framework maximizes the probability of the true kinetic rate model being retrieved from the data used, proving the essential role of having a rigorous model selection method.

AUTHOR INDEX